D0602879

"When everything around us turns to chaos, we seek security in what feels comfortable and familiar. Thus, the world has honored—and justly so—the strong, brave men who came to the rescue on September 11. Yet the world has too often forgotten a less-familiar face of heroism: that of the strong, brave women who came to the rescue as well. No tribute to the dead or living of September 11 is complete without them."

—Terese M. Floren, executive director, Women in the Fire Service, Inc.

"Ground Zero holds too many stories of bravery and sacrifice. But as a society, it is important for us to hear just a few more—about the women at Ground Zero. They, too, were valiant, and until now, their stories have been largely untold. Their valor serves as testament to their contributions and their spirit."

—Margaret Moore, director, National Center for Women and Policing

WOMEN AT GROUND ZERO

WOMEN AT GROUND ZERO

Stories of Courage and Compassion

Susan Hagen and Mary Carouba

A Pearson Education Company

International Standard Book Number: 0-02-864422-0
Library of Congress Catalog Card Number: 2002109586

04 03 02 8 7 6 5 4 3 2

Interpretation of the printing code: The rightmost number of the first series of numbers is the year of the book's printing; the rightmost number of the second series of numbers is the number of the book's printing. For example, a printing code of 02-1 shows that the first printing occurred in 2002.

Printed in the United States of America

For marketing and publicity, please call: 317-581-3722

The publisher offers discounts on this book when ordered in quantity for bulk purchases and special sales.

For sales within the United States, please contact: Corporate and Government Sales, 1-800-382-3419 or corpsales@pearsontechgroup.com

Outside the United States, please contact: International Sales, 317-581-3793 or international@pearsontechgroup.com

All photos by Joyce Benna, except for pages 292, 298, 300, 308, and 314.

Publisher: Marie Butler-Knight
Product Manager: Phil Kitchel
Managing Editor: Jennifer Chisholm
Senior Acquisitions Editor: Randy Ladenheim-Gil
Development/Senior Production Editor: Christy Wagner
Copy Editor: Jan Zunkel
Designer: Doug Wilkins
Creative Director: Robin Lasek
Layout: Angela Calvert

Dedicated to the three women rescuers
who gave their lives to save others
on September 11, 2001:

Yamel Merino, Emergency Medical Technician
MetroCare Ambulance Service

Captain Kathy Mazza
Port Authority of New York and New Jersey Police Department

Moira Smith, Police Officer
New York City Police Department, 13th Precinct

We will never forget.

CONTENTS

INTRODUCTION

Like millions of other Americans, we couldn't take our eyes off the news. As we watched in horror and disbelief, televised replays showed four hijacked planes crashing into the World Trade Center in New York City, the Pentagon, and in the farmlands of rural Pennsylvania. The day was September 11, 2001. It was a day that took the lives of thousands of Americans and shook the lives of us all.

The devastation at the World Trade Center complex claimed the greatest loss of lives on that tragic day. At 8:46 A.M., American Airlines Flight 11 from Boston to Los Angeles plowed into the North Tower, erupting into a ball of flames. At 9:03 A.M., United Airlines Flight 175 from Boston to Los Angeles struck the South Tower, igniting a second inferno. Within 2 hours, both 110-story skyscrapers had been reduced to a 7-story pile of smoldering rubble. Seven World Trade, a 47-story office building adjacent to the Twin Towers, collapsed just after 5 P.M., and several other buildings in the immediate area were damaged or completely destroyed.

An estimated 25,000 people were evacuated safely from the World Trade Center complex that day, but the loss of life was staggering. As of this writing, 2,823 civilians and rescue workers have been confirmed missing or dead. More than 1.8 million tons of debris have been removed from the site, and more than 17,000 body parts are stored at the morgue, awaiting identification through DNA testing.

Among the dead and missing are 23 police officers from the New York City Police Department (NYPD), 37 police officers from the Port Authority of New York and New Jersey Police Department, 343 uniformed members of the Fire Department of New York (FDNY)—including 2 paramedics working under the FDNY's Emergency Medical Services Command—and 6 paramedics and emergency medical technicians employed by hospitals or private ambulance companies.

In the days and weeks that followed the attack, rescue and recovery efforts at what had become known as "Ground Zero" took on a fevered pitch. As rescue workers mourned the loss of their comrades and searched around the clock for survivors, New York's "firemen" and "policemen" emerged as the new American heroes. The media presented story after story about the "return of the manly man" and made daily, unapologetic references to "the brothers" and "our brave guys."

Despite the media's common use of terms that implied a connection between gender and occupation, we knew there had to be women firefighters, police officers, and emergency medical personnel on the scene of the disaster. Susan is a firefighter/emergency medical technician in rural Sonoma County, California, and Mary is an investigative social worker who interacts regularly with law enforcement personnel. We both work with numerous women in the public safety sector, and we knew that, despite the media's failure to seek them out and tell their stories, women were making noteworthy sacrifices and contributions at Ground Zero.

In fact, three women rescuers gave their lives to save others on that terrible day. Port Authority Police Department Captain Kathy Mazza and NYPD Police Officer Moira Smith were last seen inside the Twin Towers, helping to evacuate thousands of terrified people to safety. Yamel Merino, a young emergency medical technician and single mother, was also lost in the collapse. The deaths of these heroic women were largely overshadowed by public grief for the overwhelming number of male rescue workers who were killed on September 11.

What began as our rising irritation with this lack of acknowledgment grew into outrage and then developed into a focused course of action. One of the things that bothered us most about the invisibility of women at Ground Zero was that the media presented few role models for girls and young women who might be considering careers in public safety. Women have worked hard to make inroads into occupations that were previously available only to men, and we did not want the women working at Ground Zero to fade into the background of American history. Nor did we want to see current and future generations of children grow up believing that only men are strong, brave, and heroic.

A few weeks after the attack on the World Trade Center, we boarded a plane for New York City to find these women and tell their stories. As native Californians, neither of us had spent much time on the East Coast except as occasional tourists. There hadn't been time to make formal inquiries before we departed, and we didn't have a single name in hand. All we knew was that the women were there and that we would have to find them ourselves.

Shortly after our arrival in Manhattan, we received a message from a friend of Lois Mungay, the most decorated female firefighter in the FDNY. We spoke with Lois, who was part of the first group of women hired by the FDNY in 1982. Lois believed in what we were doing, and she became our entree into the United Women Firefighters in New York City. Each interview led to the next, and we ended up interviewing eight women from the FDNY and including their stories in this book.

During the course of our interviews, we were shocked to learn that these eight women made up a full one third of female firefighters in New York City. Out of 11,500 firefighters in the FDNY, only 25 are women. Of the nearly 1,000 probationary firefighters hired since September 11, only 1 was a woman.

In 1977, women were allowed to apply to the Fire Department of New York for the first time. About 500 women passed the written exam, but after a flurry of negative press questioning women's ability to do the job, only about 90 showed up for the physical portion of the exam. Not a single woman passed.

One of the applicants, Brenda Berkman, filed a federal sex-discrimination lawsuit against the City of New York, charging that the physical test—which included climbing an eight-foot wall and performing a one-arm bent-elbow hang—was not job-related. She won her lawsuit five years later, and the city developed a new test that involved dragging hoses, raising ladders, rescuing dummies, and simulating forcible entry. Close to 50 women passed muster.

In 1982, 40 women were hired by the FDNY and began their training at New York's Fire Academy. Many of these same women are veterans of the FDNY today, eligible for retirement as they approach 20 years of service. Three of them—Rocky Jones, Marianne Monahan, and Brenda Berkman—are captains, and Ella McNair is a lieutenant. It has not, however, been an easy road to travel.

Despite the challenges, these pioneering women persevered. Not only have they paved the way for women to enter fields previously unavailable to them, but their training, experience, and tenure in the fire service also positioned them to respond in equal partnership with their male counterparts to the events of September 11. They are now seasoned veterans whose impact on rescue and recovery efforts at Ground Zero is unparalleled in American history.

Not all U.S. fire departments have such a small percentage of female members. According to Women in the Fire Service, Inc., there are currently about 6,200 women in the United States serving as career firefighters and officers, and an estimated 40,000 women in volunteer fire companies.

In Minneapolis, Minnesota, 72 women are among the 460 uniformed members of the fire department. Two years ago, Eileen Lewis, the first female firefighter to join the fire department in Tacoma, Washington, was named chief of the department. Rosemary Cloud was recently named chief of the East Point Fire Department in Georgia, becoming the first African American woman ever to become chief of a career-level fire department.

The terrorist attacks on America set the stage for the first wave of women in the public safety professions to respond to a national tragedy of such magnitude. Among them were hundreds of emergency medical personnel. At least one third of the paramedics and emergency medical technicians currently serving New York City are women, including one division chief and three deputy chiefs in the Fire Department of New York's Emergency Medical Services (EMS) Command.

FDNY-EMS employs 3,000 paramedics and emergency medical technicians (EMTs), making it the largest EMS agency in the country. During our first trip to New York City, we saw three female paramedics from the FDNY briefly interviewed on an NBC *Third Watch* television special. The next day, we tracked them down. Captain Janice Olszewski, Lieutenant Amy Monroe, and paramedic Christine Mazzola all agreed to share their stories for this book. They also introduced us to two other members of FDNY-EMS—Lieutenant Kathleen Gonczi and EMS dispatcher Lieutenant Doreen Ascatigno—whose stories appear in the pages that follow.

Because EMS personnel wear the FDNY logo and work from FDNY ambulances, they are recognized as members of the Fire Department of New York. But some 39 percent of 911 ambulances in New York City are operated by hospitals. These and other "voluntary" agencies—including volunteers and private companies—lost six paramedics and EMTs in the collapse of the World Trade Center, including 24-year-old EMT Yamel Merino.

These casualties have often been overlooked in the death toll of rescue workers reported by the media. Because their loved ones were not considered municipal employees, the families of these victims were not eligible for the same benefits received by those of NYPD and FDNY personnel who perished in the same incident.

In the course of our investigation, we learned that the general public is largely unaware that these agencies exist as entities separate from the FDNY. We interviewed two EMTs from voluntary hospitals—Mercedes Rivera and Bonnie Giebfried—who told us that the invisibility of their colleagues at the tragic events of September 11 has only made it more difficult to reconcile the loss of friends and co-workers.

The stories about women in EMS are particularly harrowing, because the usual role of emergency medical personnel in a Mass Casualty Incident is to establish a triage center in a safe area to treat and transport patients who are brought out of the danger zone by firefighters. EMS workers are not equipped with breathing apparatus or protective clothing, nor are they trained to attempt rescues under hazardous conditions.

The events of September 11 altered their normal means of medical practice by placing them inside the danger zone. The fires and rubble of Ground Zero were spread across 17 acres, and the sheer magnitude of the incident made for unprecedented personal and professional challenges.

Members of the New York City Police Department were also faced with challenges never before encountered on such a massive scale as they responded under the highest level of police alert. Of the NYPD's 38,000 uniformed officers, more than 6,000 are women.

We found NYPD Sergeant Carey Policastro through the National Center for Women and Policing. Carey was immediately drawn to the concept of this book. In addition to sharing her own story with us, she put us in touch with several other female police officers who responded to the World Trade Center on September 11 and in the days that followed. Again, one interview led to the next until we had collected 10 stories.

Many of the police officers we interviewed told us about their repeated brushes with death—and injuries they sustained—as they dodged falling debris while evacuating thousands of civilians. Detective Jennifer Abramowitz told us about raking through debris at the landfill on Staten Island, where the rubble from the Twin Towers was ultimately delivered by barge. Police Officer Maureen Brown described the horrors of her assignment at the morgue, where hundreds of bodies and body parts were delivered every day.

While the media has focused predominantly on members of the NYPD and the FDNY in its portrayal of the new American heroes, the Port Authority of New York and New Jersey Police Department has been scarcely recognized for its role in the events of September 11. Most people are unaware that the World Trade Center was a Port Authority facility. The Port Authority, which maintains and controls all the bridges, tunnels, trains, and buses that connect New York and New Jersey, employs about 1,350 police officers, 99 of whom are women.

We interviewed Port Authority Police Officer Sue Keane, who credits her military and police training with saving her life as building after building disintegrated around her. Sue survived the attack but mourns the loss of 37 Port Authority police officers—including her colleague, Captain Kathy Mazza—in the World Trade Center disaster.

On September 11, the Port Authority Police Department suffered the largest single-day loss of any police department in the history of the United States.

An unprecedented number of women responded to the events at the World Trade Center. They came from every Borough in New York City and from all across the country as firefighters, police officers, emergency medical providers, military personnel, and members of search-and-rescue teams. Women in the trades contributed their skills as construction workers, heavy equipment operators, steelworkers, electricians, utility workers, and engineers.

Women were also among the first to arrive at Ground Zero to organize relief efforts for the American Red Cross, the Salvation Army, the Federal Emergency Management Agency, and other organizations. Female doctors, nurses, counselors, physical therapists, and massage therapists brought the healing arts to ailing rescue workers on the scene. Women volunteered to do anything and everything they could to help.

While this book is largely composed of the stories of rescue workers, we've included interviews with two counselors—police psychologist Sarah Hallett, Ph.D., and Major Molly Shotzberger of the Salvation Army—who tended to the emotional needs of rescue workers. We also interviewed Major Kally Eastman, an engineer with the U.S. Army Corps of Engineers, and Rose Arce, a producer with CNN, each of whom contributed to the rescue and recovery efforts at Ground Zero in her own unique way.

This book is not meant to diminish the contributions of the men who lost or put their lives at risk at Ground Zero. It is not meant to overlook the suffering of the families of the countless civilians and rescue workers who were injured or buried under the rubble. Nor is it meant to minimize the stories of heartbreak and heroism that emerged from the Pentagon or the fields of Pennsylvania on that fateful day. It is simply meant to tell the story of the World Trade Center tragedy from a perspective that has been largely ignored—that of the women who were there.

Walking the desolate streets of Lower Manhattan weeks after the attack, we were painfully aware that everyone—from the cabbie who ferried us across town to the deli worker who built our sandwiches—had a story of worry, loss, and fear associated with September 11. As we neared the perimeters of Ground Zero, we were compelled to talk with every female police officer we saw on post behind a barricade. We knew we couldn't interview everyone, but each time we passed a woman in uniform and didn't stop to talk, we felt guilty for leaving her out of this book.

The interviews we did conduct were long and tiring and emotional. Almost everyone cried, including us. Many of the women we interviewed told us that they'd been so busy working, they hadn't stopped to tell their stories from beginning to end. Sitting across the table from us was the first opportunity they'd had to slow down and remember the details. It was a devastating recollection of memories for most of them, although many told us that they felt better afterward.

We saved the families for last. The last three stories in this book are told by the families and friends of the three women who perished in their quest to save others. We waited to interview them until the final hours before this book went into production, because we wanted to give them time to adjust to lives without their loved ones—lives that will never be the same.

We experienced so much heartbreak during these last interviews. The families who shared their fresh sorrow with us reminded us that there are thousands of other stories as heartbreaking as these tied to the other rescue workers and civilians who died on September 11. Our moods remained somber in the aftermath of these final interviews as the book made its way to press.

With the publication of this book, the stories of the women at Ground Zero leave our hands—though never our hearts—and are passed on to you, the reader. It is our hope that the contributions and achievements of such women will find their way into the annals of American history and into the story lines of the mainstream media. Most important, we hope that these stories will inspire our nation's young women to follow their dreams and believe in the potential for their own lives.

ACKNOWLEDGMENTS

We are grateful to the women whose stories appear in this book. We thank them for inviting us into their homes, their places of work, and their personal and professional lives. We are humbled by their honesty and inspired by their courage. We honor them for committing their lives to public service and setting an example for young women everywhere.

We are grateful to the families and friends of Yamel Merino, Kathy Mazza, and Moira Smith for talking with us about the loss of their wives, mothers, daughters, sisters, co-workers, and friends. We know that our inquiries opened up fresh wounds that had scarcely begun to heal. We thank them for trusting us with their memories and allowing us to share them in this book.

We also wish to thank the women whose stories do not appear within these pages. During the early stages of our research, we spoke to a number of women whose experiences on September 11 helped to guide and inform us as we developed the content and theme for this book. We especially wish to thank Sheila Frayne, Angie Bucknell, Margo Currie, Kim Grosso, Noreen Dean Dresser, and Amanda Schmidt.

There are many others who wanted to include their stories in this collection, but whose work schedules or employers would not allow the time or permission. To every woman who responded to the World Trade Center tragedy on the day of the event or in its aftermath, we offer our gratitude for their selfless actions and the contributions they continue to make in the work they do every day.

We are also grateful to the many men who told us about their women colleagues, helped us to locate and schedule time with them, and offered other invaluable assistance along the way. The pride with which they described their female co-workers was inspiring and very much appreciated.

Our photographer, Joyce Benna, is also worthy of our thanks. Not only are we grateful for her technical skill as a photographer, but also for the creativity, flexibility, and good humor she brought to the job. Joyce endured long days of shooting, unpredictable weather, and last-minute schedule changes with grace and aplomb. More than once she scrambled to move her equipment out of the way of fire apparatus when an alarm came into a firehouse where we were shooting. Never did her enthusiasm for this project wane.

We are grateful to our literary agent, Jessica Faust of BookEnds, LLC, for believing in this book from its inception. We thank Randy Ladenheim-Gil and Christy Wagner, our editors at Alpha Books, for making our first book-publishing experience such a joy. We also wish to acknowledge our good friend and fellow author, Suzanne Sherman, for her invaluable assistance.

Finally, we wish to thank the Hagen and Carouba families, the County of Sonoma Human Services Department, the firefighters of the Graton Fire Protection District, our friends, and our greater community in Sonoma County, California, for their tremendous support of this project. Every person who offered encouragement, financial assistance, or simply a kind word contributed to making this book possible.

NYPD
NYPD 9-11
9832

CAROL PAUKNER

AGE 37
POLICE OFFICER
NEW YORK CITY POLICE DEPARTMENT
TRANSIT DIVISION, DISTRICT 2
LOWER MANHATTAN

It was Carol Paukner's quick wit and easy laugh that made her so engaging, and her big heart that made us want to stay in touch. Twice when we returned to New York City to finish the photography for this book, Carol volunteered to drive us from one appointment to the next on her days off. She even took us to her home in Long Island to have dinner with her family before delivering us back to the train station.

"What happened on September 11 has given me a new perspective on what life's all about," she told us over chicken cutlet parmesan and baked clams. "We're putting in this whole new kitchen because, why wait? What are we waiting for? Instead of putting things off until tomorrow, I'm dealing with them today."

Carol still had respiratory problems and was recovering from shoulder surgery the last time we spoke in early spring 2002. "Once my shoulder heals," she said, "I'll go in and have my knee operated on." Her injuries are a daily reminder of what she experienced that tragic day.

Even though 25,000 people were evacuated from the World Trade Center on the morning of September 11, Carol said she still feels defeated because so many people died.

"I keep thinking that if I'd done something differently," she said, "maybe more people could have been saved."

"How could I leave all these people? I said to another officer, 'I'm going to die today.' I just looked at the building and said, 'This is my day.'"

—Carol Paukner, November 24, 2001

"I've been patrolling the New York subways for about 10 years, with the exception of a year in Narcotics. We're usually given one or two stations to cover. On September 11, my assignment was the Broadway Nassau subway station, which is one block away from the World Trade Center.

"That morning, a call came over the radio for an unknown condition at the World Trade Center. My rookie partner, Tracy Donahoo, and I went upstairs, and we saw the streets covered with debris. People were running, yelling, shouting, crying. I've never seen anything like that before in my life. Some woman said, 'It's over there!' and pointed toward the World Trade Center. We ran up the road, and as I looked up, I saw the first plane half in, half out of the North Tower. I only saw the back end of the plane, the tail end of it. The nose was buried in the building. I thought I was in a dream. There was debris everywhere. I grabbed my radio and confirmed, 'A plane has hit the World Trade Center.'

"I didn't think this was real. The plane didn't look like it was very big because it was so far up and there was such a small portion of it I could actually see to identify it as a plane. We ran to the base of the second building where the bookstore was located and immediately started evacuating people. They were running through the doors, and we were trying to calm them down. Meanwhile, there was stuff falling off the building, and we tried to keep everybody under the overhang. We were telling them, 'Please stay under the shelter! Move to the left, go up Church Street, and get out of here!'

"A lot of people were distraught. A lot of them wanted to look up and see, but we were telling them, 'Please leave. You have to leave.' It's human nature. People want to stop. They want to see. They want to say, 'Hey, look what happened! I can't believe it! Oh my God, a plane!' Meanwhile, I know what's happened, and I'm trying to tell them, 'You have to leave. Get out of here. Go.' A lot of them did listen to me, and I assisted many people who were injured out of the building. One woman had multiple sclerosis, and I carried her up the escalator and carried her up to Broadway, one block away, where they were setting up

The Story of Women at Ground Zero

In response to the media's portrayal of rescue workers as "firemen, policemen, brothers, and our brave guys," Northern California authors Susan Hagen and Mary Carouba traveled to New York City to find and interview female firefighters, police officers, paramedics, and other women rescuers who responded to the World Trade Center tragedy.

The two West Coast writers pooled their credit cards and set out for Manhattan just a few weeks after the attack on the World Trade Center. Despite the fact that they had no contacts, no press credentials, and knew no one in New York City, they were determined to find the women rescuers and tell their stories. The result is *Women at Ground Zero,* a collection of vivid and emotional stories told to them by the women who risked their lives to save others on that tragic day.

"One of the things that bothered us about the invisibility of women at Ground Zero was that the media presented few role models for girls and young women who might be considering careers in public safety," Hagen and Carouba write in the book's introduction. "Women have worked hard to make inroads into occupations that were previously available only to men, and we did not want the women working at Ground Zero to fade into the background of American history. Nor did we want to see current and future generations of children grow up believing that only men are strong, brave, and heroic."

Susan Hagen is a former news journalist and retired firefighter/EMT. Mary Carouba is an investigative social worker and advocate for abused children and elders. Since the publication of their book, they have been interviewed on dozens of news and talk shows across the country and have received widespread media coverage for their insightful book. Mary and Susan continue to make personal appearances throughout the country to inspire others to follow their dreams.

Mary Carouba Susan Hagen

Susan and Mary's presentation begins with a candid and surprisingly humorous look at their inspirational journey, includes vivid and emotional stories told to them by the women who risked their lives at the World Trade Center, and concludes with a moving tribute to women rescue workers who made the ultimate sacrifice to save others. Their talk includes an audio-visual presentation with photos taken by rescue workers at the scene on 9/11, as well as portraits and quotes from the women whose stories appear in the book.

The powerful stories of the women rescue workers of 9/11, coupled with the subtext of two unpublished authors whose passion and determination caught the attention of a major publisher, underscore the message that anything is possible. Many audience members have remarked that they leave Mary and Susan's talks believing that the possibilities for their own lives are indeed limitless.

"Susan and Mary's presentation to the American Association of University Women at our California State Convention inspired and energized the attendees. Their wit and insight provided everyone with a strong sense of the power of women and the importance of their contributions at Ground Zero. Susan and Mary are to be commended for making sure that these stories are recorded in our history."

Sue Cochran, President
AAUW California

"The Women at Ground Zero presentation was not only inspirational, but provided invaluable information for our law enforcement executives. Susan and Mary's sense of humor and easy, comfortable speaking style made their talk seem intimate and personal, despite the large size of the audience. Many people remarked that these were the best speakers we've had at any of our conferences."

Brenda Van Amburg
National Association of
Women Law Enforcement Executives

ABOUT THE BOOK:

"Women at Ground Zero is a moving book that adds significantly to the stories of bravery spurred by the tragedy."

New York Daily News

"The interviews convey the triumph of the human spirit despite adversity. And they weave a tragic, unfathomable tale packed with gripping first-person dialogue."

The Chicago Tribune

"This praiseworthy book comprises interviews with women rescue workers. David Halberstam's Firehouse and Richard Picciotto's Last Man Down contain few women. Stoicism and ellipsis mark those best-selling books. By contrast, these women divulge. They summon details. The women's capacity for wonder and doubt -- and their willingness to express it -- fills the book."

The Washington Post

Women at Ground Zero Order Form

To receive a copy of the book signed by both authors, send $30 for media rate (allow up to 4 weeks for delivery) or $35 for first-class shipping (allow 10 days for delivery):

☐ I am enclosing a check for $______ (Please make checks payable to Women at Ground Zero)
☐ Please charge to Visa / MC $______ Card Number:______ Exp. Date______ Signature______

Name: ______ Address: ______

City: ______ State: ______ Zip: ______ Phone: ______

☐ Please notify me of Women at Ground Zero events via my **email address:** ______
Indicate any special requests regarding signing — to whom book should be inscribed, special message, etc.: ______

______

Detach and mail this panel along with your check to: Women at Ground Zero, PO Box 506, Sebastopol, CA 95473 - (707) 824-1529

"To all the people who said, "Thank you" to us, I want to say 'Thank you' to them. It kept our hearts going when we were feeling nothing."

Patty Lucci,
Police Officer, NYPD

100% of author profits from the sale of the book *Women at Ground Zero* will be donated to Women in the Fire Service, Inc., and the National Center for Women and Policing.

To learn more about the heroic women of 9/11, leave messages for them in the Women at Ground Zero Guest Book, or view a listing of the author's speaking engagements, please visit our website at **www.womenatgroundzero.com.**

Women at Ground Zero
Post Office Box 506
Sebastopol, CA 95473
(707) 824-1529
carouba@comcast.net
suzhagen@sonic.net

"I think the essence of love is laying down your life for others. If you've ever felt that nobody loves you, just look at a fire truck going by. Then you'll know that somebody does."

Regina Wilson,
Firefighter,
FDNY

a triage center. There was a big, tall gentleman in Army fatigues, and I was so tired and out of breath from carrying her that I said, 'Sir, would you please assist me and carry her the rest of the way around the corner?' He said, 'Surely,' and he helped her out.

"At that time, I knew there were more planes coming. The FBI and the Federal Marshals were there, and they said, 'You're not a coward if you want to leave. There are more planes coming. This is terrorism.'

"But how could I leave all these people? I said to another officer, 'I'm going to die today.' I just looked at the building and said, 'This is my day,' and I went back and continued to try to evacuate people without causing them any more alarm. I said to my partner, 'If anything happens, if and when another plane comes and hits the building and we get separated, meet me in that corner over there. I'm going to dive for that corner.'

"People were coming out of the building, and they were totally lost. They were saying, 'What do I do now? Where do I go?' I told one person, 'Hey, this woman needs some help; just go with her, stay with her, and keep heading up Church Street.' I was trying to give them some support so they would stay together. I knew that once they got out there, if they did look up, these people were going to freak out.

"Then the jet fuel from the first plane came down the elevator shaft and caused a big explosion. With that explosion, more people came running out of the building. Everybody was pushing and shoving and screaming. One piece of metal fell to the ground two feet away from me. It was the size of a car hood. It hit so hard that the impact blew me into the glass partition of the bookstore and knocked me to the ground.

"Two cops were yelling at me, 'Get up! Run! Get up! Run!' I wasn't even thinking I was hurt. I wasn't feeling anything. I was out there to survive. I was able to get up and run, and I got into a little doorway with them. One of them was Nancy Ramos, a rookie with the Manhattan Traffic Task Force. We stayed there until the explosions stopped, until it was almost silent again, then we all ran back and continued evacuating people from the building. It seemed like we were there for hours. My adrenaline was just going and going and going.

"At that point, about 15 firefighters arrived to assist in evacuating the building. They were in full gear, carrying ropes, packs, axes, and hoses. I made way for them, cleared a path through the people, and got them into the building. It bothers me now to think that those men went into the building and I don't know if they came out alive. I don't know their names or the company they were with, so I can't check on them. It really bothers me that all those guys who walked past me probably died.

"About 20 minutes later, we heard a huge plane coming in. We couldn't see anything because we were under the overhang, but we heard this plane and the echo of it. Then it hit the building. I got blown through the exit, but I was able to catch the door of the building as I came out. I just held on. There was so much smoke and soot that I couldn't breathe. I was thinking, *This is it. I'm going to die.* I was half in, half out of the building. People were blowing past me, particles were flying, people were flying. Stuff was coming down right on top of me, and I couldn't see anything. I held on with one hand, and the wind force was so strong that I couldn't get my other hand up to the doorway to pull myself through.

"Then my hand hit a leg. I pulled on the leg, and it was a man. He was alive, and he was screaming to me because it was so loud. 'Grab my hand! Grab my hand!' So I grabbed his hand, I pulled myself to him, and we huddled in this corner that I originally thought looked like a safe place, where I'd planned to meet my partner. I just covered up with him and held on to him. We were lying on bodies and trying to hold on.

"Then it was silent. We couldn't breathe, we were choking, and we couldn't see because of all the stuff in our eyes. I didn't know who this man was. I didn't know if he was a firefighter, if he was a cop, or if he was a civilian. He said to me, 'Do you have a flashlight?' I said, 'Yes, I do,' and I pulled the flashlight out of my belt.

"But even with the flashlight, the smoke was so thick, we still couldn't see. He said, 'Don't let go of me.' I said, 'I'm not letting go of you,' and he said, 'I'm not letting go of you, either.' We were in there alone, we were trying to crawl out, and we were calling for people, but a lot of them were already dead. We continued to crawl, but there was so much debris that we couldn't crawl very far, so we stood up and felt our way around the building. Finally he said, 'I know where I'm going.' I said, 'Wherever you're going, I'm going with you. If I'm going to die, I'm not going to die alone.'

"Meanwhile, we were throwing up, and we were trying to clear all the dust and debris off ourselves. We didn't know what side of the building we were on or where we were, because we'd been knocked around so much. Through all of this, we kept hearing a man's voice saying, 'Holy Mary, Mother of God.' Then he'd say, 'Jesus Christ.' He kept repeating the same lines, over and over again. Later I asked the guy I was with, 'Did you ever physically see this man?' He said, 'No.' I said, 'Did he ever physically touch you?' He said, 'No.' But we both heard him talking to us. My dad had just died, and I'd like to think it was my dad who was there with us, who talked us out.

"Then we saw trees, and we realized we were outside. It was so black that we still couldn't see, and all we heard was this 'beep, beep, beep, beep.' At the time, we thought it was car alarms. I didn't know until later that it was the sound of the alarms that go off when firefighters are down and there's no movement.

"When we got to the street area, we saw all the crushed fire trucks and ambulances and dead bodies—parts of people—pieces that looked like birds. By then the air was starting to clear a little, and I saw that the man I was with, Richie Vitale, was a cop. We were flipping out, holding each other, and saying, 'You saved my life! You saved my life!' We didn't know where our partners were, and we were very much in shock. I had lost my rookie partner, and he had lost two of the rookies he had been with. One of them was Nancy Ramos.

"We knew we had to get away from the area, that we weren't safe there, and we both still felt that we were going to die. There were so many casualties, I thought we were going to die from what we had inhaled. I kept vomiting, and I thought the smoke was going to kill us because there were so many people lying there dead already. I thought that, eventually, I was going to drop, too.

"We tried to help anybody we saw who was still alive. There was a lady standing there, holding her hands open, looking up and saying, 'I lost my pocketbook. I lost my pocketbook.' We grabbed her and said, 'Ma'am, we have to get out of here. Come with us.' We were grabbing other people who were lying there and trying to help them, and all the while this lady kept looking up with her hands out, saying, 'I lost my pocketbook. I lost my pocketbook.' I was reminded of that famous picture of the little girl covered in Napalm, running down the street in Vietnam.

"We came to a deli, and we went in and took water and tried to clear off our faces, clear our lungs, and clean out our eyes. We both had scratched and burned corneas. We grabbed bottles of water and walked about a block from the deli, then another police officer who knew Richie called us into a building where he thought we would be safe. As soon as we went in, Richie fell to the floor. He couldn't breathe. I thought he was having a heart attack. I started yelling at him, 'You can't have a heart attack on me now! Are you kidding me?' I started punching his arm. I was like, 'What, are you crazy? After all that, you're not going to die on me now!' The other guys were saying, 'Take it easy. Take it easy.' I was screaming, 'No! You don't understand!'

"Someone was there with oxygen and gave Richie medical treatment, and that's when the first building fell. Black smoke started rushing down the street.

The cars were popping, and everything was on fire. We moved out of the entrance and went to the back of the building, where there was a big garage with aluminum doors. There were a bunch of firefighters there, and they took a hose inside and were hosing it down to prevent more smoke from coming in. But the smoke was so thick that we still couldn't breathe. We had to get out of there.

"I didn't know if I was going to live or die, breathing in all of those chemicals. Was it biological warfare? At that point, seeing the streets full of ash, and with everything that had happened there, I was just in shock. Total shock.

"We started to walk toward the West Side Highway and met up with Richie's lieutenant. I told him, 'I don't know where my partner is. Please call Transit District 2 and let them know she's missing but I'm okay.' He said, 'You're in my care now. Just stay with me.' He called in and made a notification that I would be staying with him, and we continued on toward the West Side Highway, helping other firefighters and people who were injured along the way.

"When we got to the West Side Highway, we finally thought we were safe. Richie and I sat down on a bench, and we were crying and holding each other. Some cops came by and said, 'You have to leave the area. There's a gas leak.' We got up and started walking again, and one of those police trucks that carry barricades came down the street. It was empty, going down the highway with nobody on it, so I knocked on the guy's side door and he stopped because he thought he had hit somebody. I said, 'Let us on the truck!' So we climbed up into the back and pulled four firefighters, some EMS guys, and more cops up with us. I said, 'Now you go. The truck is full.' He took us to a triage center.

"I ended up going to St. Clare's Hospital in the back of an ambulance with six men. One was a firefighter, and the rest were cops. When we finally got to St. Clare's, they had no oxygen left, so we were sharing oxygen bottles. The doctors said, 'We really want to admit all of you, but we can't, because we're going to have so many other people to admit.' I said, 'No, you're not. They're all dead.'

"We were released from the hospital early that evening, and we went to Richie's precinct and sat and watched it all on TV. We couldn't believe we had been there. We couldn't believe we were alive. We couldn't believe how many people had died. But we also knew that we had saved a lot of people. A lot of people got out of the buildings. My partner was found, and she was okay. Richie's two partners were okay, too.

"Richie and I still talk, and when we have our hard times, he'll give me a call or I'll give him a call. We visit each other, and we know each other's families. I feel very close to him. Some days it can be difficult. The day after, I heard a very loud plane and I actually ran out of my house because I was so afraid. After living through this, it's something I'll never forget. Having a building fall on me, seeing so much death, dealing with this tragedy—it's a lot for anybody to take in.

"I would love to work out with my weights to relieve the stress, but I can't. I can't lift more than five pounds right now, because I have a torn rotator cuff in my shoulder, and my knee is torn in three places. I'm going to need surgery on both. I also threw my neck out, hurt my right foot, and burned my eyes. I still have a lung infection. I feel like somebody's sitting on my chest. I can't get rid of this cough or sleep lying down. I've been treated for it, but it won't go away.

"So that's a difficult thing. I'm on restricted duty with the police department because of my injuries. Now I'm answering phones and going to physical therapy. I would love to get on a bike or push some weights to release the tension, but I want to do the proper thing to heal. I talk a lot. I have a good support system, good friends, a good spouse. I'm not a big drinker so, thank God, I haven't done that. I get up, I go to work, and my life continues. I have children in my life, and they keep my mind occupied. I try to keep busy. I try not to think about it.

"There's not one cop I know, or one firefighter, who hasn't thought about leaving the job, who hasn't said, 'Hey, what am I doing here?' But I think going back on patrol would be a good thing for me. I always wanted to be a cop, all my life, since I was a little girl. I didn't want Barbie dolls. I wanted police action figures. I have that in me. I'm able to help people, and that's what I did that day. I know I did my best, but I wish I could have done more.

"I spent 10 years patrolling the subways in that area of Lower Manhattan. The Port Authority Police Department had its headquarters in the World Trade Center, and as a police officer I would often go into their headquarters to take a break, use the facilities, or have a meal. At some point in time, I probably ran into all the Port Authority officers who were stationed there. I'd say, 'Hi, Officer, how you doing? Hi John, hi Bob.' I didn't know them all that well, but we'd sit there and have lunch together. Now they're all gone. They all died.

"And on the concourse level, where all the shops were, I'd go in and have a slice of pizza for lunch or just walk past the newsstand and wave to the guy, like, 'Hey, how you doing? What's going on?' It's weird for me, because there's no place for me to go back to and say, 'Are you okay? Did you make it out all right? Is your family safe?' I can't go back and walk through the strip and say, 'Hey, the newsstand man! He's okay!' Or the perfume shop guy. Or the people in the Warner Brothers store. I walked that strip all those years, and I got to know these people by their faces. I'll never know if that gentleman at the deli who served me soup once a week got out of the building alive."

35
12077

MAUREEN McARDLE-SCHULMAN

AGE 42
FIREFIGHTER
FIRE DEPARTMENT OF NEW YORK
ENGINE 35
HARLEM

It was barely 6 A.M. on a chilly morning in November when we met up with firefighter Maureen McArdle-Schulman at the New York Fire Academy. It was the first week of training for a new class of probationary firefighters, and Maureen had just begun the first of two eight-week details as a training officer.

"I don't think there's anything more rewarding than putting out a fire," she said. "You go in, you do what you need to do, and you're done. That's it. It's a wonderful feeling of accomplishment."

Maureen should know. With 19 years on the job as a firefighter with the Fire Department of New York, she's long since lost track of the number of fires she's helped extinguish. After offering her experience and background to help train the first class of firefighters to enter the Academy since 343 members of the department were lost on September 11, Maureen returned to Engine 35 in Harlem.

"To admit that my life has changed since that day would give those evil people power over me," she said. "It's been said that the best thing to do is to get back to doing your normal things. Maybe getting back to normal things is postponing some of the feelings about what happened on September 11, but I've got to do what I've got to do. The world can't just stop."

"What we were trained to do and what we were able to do that day were two completely different things."

—Maureen McArdle-Schulman, November 6, 2001

"It was just a fluke that I picked Tuesday, September 11, as one of the days to do my mandatory overtime. I have three kids and my husband works nights, so it's very difficult to get the overtime in. I was actually on vacation that week, but I scheduled myself to come in that day because my husband works Tuesday days and we'd be home together that night.

"I've been here in this same firehouse in Harlem for 19 years, and the locals are still saying, 'Oh my God, there's a woman on the rig!' Maybe it has to do with the fact that I put my hair up when I'm working and I'm in bunker gear most of the time, so they haven't always noticed that I'm a woman. I've been a driver, or chauffeur, for the last six months, which means I'm the engineer on the rig. I drive it, I pump it, and I'm responsible for all the equipment. Since I've been driving, I'm not always in bunker gear and I'm not wearing my helmet all the time. So people notice I'm a woman and say, 'Wow. Where you'd come from?'

"I got to work that morning at 7:30, as usual. I changed my clothes and was hanging out in the TV room on the second floor when someone yelled to turn the TV on. That's the first we saw of what the World Trade Center looked like, just after the first Tower was hit. At that time, the filming wasn't very good, and we couldn't tell how big the plane was. It was a beautiful, clear day, and all I kept thinking was that somebody really screwed up a flying lesson.

"Because I was the overtime person, I was supposed to be detailed out to another firehouse. I ran out, got in the car, and started driving to Engine 91, which is a single-engine house on 111th Street in Harlem. I was listening to the car radio, and somebody had called in who lived right by the World Trade Center and was explaining what it looked like. Then all of a sudden—and this was so eerie—he said, 'There's another plane coming in really close!' and he described the second plane hitting the other Tower.

"I was listening to this and thinking, *This is urgent. This is really not good news.* I pulled into Engine 91, and as soon as I came in, the dispatcher announced over the loudspeaker that it was a fifth alarm. That meant a massive number of firefighters were responding. With a fifth-alarm assignment, many of the companies are on the scene for relief purposes. You stand on the street waiting for companies to come out, then you go in to relieve them. I stuck my cell phone in my pocket, got on my gear, grabbed a radio, and off we went.

"We drove through Central Park, and at the south end of the park, we picked up other companies on their way. It was a big caravan of police cars, unmarked cars, fire engines, and fire trucks—all going to the World Trade Center. They actually opened a lane on the West Side Highway specifically for emergency vehicles, so we got there in no time. We got off the rig, put the rest of our gear on, and grabbed our hose roll-up. I had the radio and the standpipe kit, which are the tools you need to access water in a building. I think we were two blocks away. We started heading down to the World Trade Center.

"I hate stairs. That's one of the reasons I stay in Harlem. We have little buildings—33 stories is the biggest one we have. In a typical high-rise fire, we would have used the elevators to probably five floors below the fire, then walked up the rest of the way. But the Tower elevators had jet fuel in them, so they weren't being used. People were taking the stairs. They said it took more than an hour for people to walk down 80 flights, so imagine how long it would take us to walk up with all our equipment and gear. The idea of climbing up all those stairs—I was getting sick at the thought of it.

"There was a Command Center set up at the mouth of a parking garage across the street from the World Trade Center. Our lieutenant went up to the table and reported in, saying that Engine 91 was present. There were about 100 firefighters already in the parking garage, awaiting their assignments. These were all the companies from Brooklyn who had come ahead of us. A five-alarm fire brought in people from all over the area, so we were seeing companies we don't normally see. It's very rare to see Brooklyn companies up in Harlem.

"All of a sudden, someone yelled that he thought part of the plane was coming out of the Tower. Well, it wasn't part of the plane. It turned out to be a person. People were starting to jump. After the first person jumped or was thrown out, many others followed. I was standing there, staring up, and I couldn't believe what I was looking at. I was thinking, *How bad could it be up there?* I felt so useless. I wanted to do something. But then I got nauseated and I stopped looking. I turned back to face the wall, but I could still hear them hitting the ground and hitting the building, because the building is wider at the bottom. I could still hear them hitting.

"Then my lieutenant was called up to the table at the Command Center and the officer said, 'Listen, we want your company to go into Tower Two, down to sub-basement six. They're having a problem with the fire pumps, and we want you to check them out.' Sub-basement six is six floors below ground level. I quickly called my husband and left a message on his answering machine at work. I said, 'I'm going into

Tower Two, sub-basement six.' Then the chief said to my lieutenant, 'You'll need forcible entry tools, because there's nobody down there to get you in.'

"Someone ran back to get the tools, and all of a sudden Tower Two started puffing and jumping. I don't know how else to explain it. There was a ring of fire around the roof, and the building above the area that was burning just started jumping up and down. Then someone came over to the table and said, 'A firefighter just got hit by a jumper. We need last rites.' So a couple guys ran to the right, and we were still standing there staring at the building. It was mesmerizing. People were still jumping from Tower One, and Tower Two was puffing and jumping. Then someone yelled, 'Run!'

"If we'd had our forcible entry tools with us, we would have been in the building already. I came out of my trance, turned around, and started running. All I kept thinking as I was running into the garage was, *I'm going to get crushed.*

"I moved over to the curb and ran my foot along it as I was going into the garage so I'd be able to find my way back out again. I still had my hose roll-up on my shoulder and the standpipe kit, and that's when the building imploded. Suddenly, it was completely dark. I had my airpack on my back, and the air became very thick and hard to breathe, so I put on my facemask. But my facemask was full of dust and all that crap went in my eyes and my throat and my mouth.

"We were in the garage, and everything was coming in because we were right across the street from the World Trade Center. It was complete darkness. Finally, someone started asking, and then everybody started asking, 'Is everyone okay?' We started feeling around the ground to make sure nobody was hurt, crushed, or whatever. We didn't know if the front of the garage had collapsed from the explosion or if the debris from Tower Two had trapped us inside.

"We felt around and found that nobody was hurt. Then a guy said, 'I know the way out.' We were all holding on to each other's sleeves, and he took us out. Even after we got out, I really thought I was still in the garage because it was so black outside. The guy next to me started having an asthma attack and asked for my face piece, so I gave it to him. Someone put him in a police van that had the air conditioning on, because the guy couldn't breathe.

"I finally got my bearings, and I went back in the garage looking for my company. I started yelling for them. '91 Engine! 91 Engine!' I found my lieutenant and the probie who was working that day, a guy with less than five years on the job. But we were still missing Timmy Hoppey. He's got a 12-year-old son who has cancer, and the kid's been through chemotherapy, radiation, surgeries, and everything else in the past year. All I kept thinking was, *We can't lose this guy.*

"The lieutenant told us to go back and stay with the rig, and when we got there, it had two inches of dust on it. I said to the probie, 'Let's move the rig. Let's get it farther away.' So we moved it a block away, got near a hydrant, and made sure we'd have water in case we needed it. I was just doing what my training had taught me to do. The sun finally came out and there was air, and I said, 'Oh. There's still a God here. There's somebody in control of this whole thing.' Before that, it was like midnight in the middle of the morning.

"Then I thought, *Oh God, I'd better call my husband. He still thinks I'm in Tower Two.* I couldn't get a signal for my cell phone to work, so I found a pay phone, but it had no dial tone either. Finally, I got my cell phone to work if I stood 10 feet from the pay phone. I called my husband and said 'I'm okay.' Then I called my father. He's a pisser, my father. He worked 25 years as a firefighter in Harlem during the 1960s and 1970s, so anything I tell him, like 'Hey Dad, got a second alarm on our hands,' he's like, 'Eh, it's nothin'. You wouldn't believe what we used to do.' You can't impress the man. I called him up and said, 'Dad, I'm down at the World Trade Center.' He goes, 'Oh, that's nice.' In the middle of all this, I was hysterical laughing on my cell phone.

"I have two brothers on the fire department. My older brother, Kevin, is in Squad 41, which lost a lot of guys, and my other brother, Chris, is in a truck company in Queens. I was worried about them. My dad told me that neither one of them had duty that day, so I was like *Phew!* At least we did okay. There was a possibility that my parents could have lost half of their kids, because there are six of us in the family, and three of us are on the fire department. So it was close.

"I got that all settled, and I went back to the rig. The other guys and I wanted to go look for Hoppey. There were millions of guys in the street, clusters of them all over the place. Everybody was covered in soot, and my eyes were on fire from all the dust and everything. We went walking down the street looking for Hoppey, and this cop said to us, 'We've got a report of a sniper in Army fatigues with an automatic rifle, so be careful.' There were gas leaks everywhere and bomb scares. We were running from one side of the street to the other, not knowing where to go. We ended up down at the pier until the gas leaks were shut down and there was a semblance of sanity again.

"We went back to the rig, and the probie and I were staring at the fire in Tower One, when all of a sudden it started puffing and jumping like Tower Two did before it collapsed. I said, 'It's going to go! That's exactly what the other one looked like before it went!' He ran across the street and dove under a rig, and I got up on the tailboard of our rig and curled up into a fetal position and just covered myself as much as I could. I was cursing myself because I had put my airpack away on the rig. I should have kept it on, but who knew? And then Tower One imploded.

"This time it was worse, because I saw this building fall. When Tower Two collapsed, I turned my back on it and ran. This one I actually saw fall. I saw it happen. I saw it coming down. It was just like a storm coming down the street. I was on the back of the rig, so I was protected by the hose bed and the rig itself. But I kept thinking, *I hope this isn't enough to blow the rig over on me.*

"With everything I did that day, I didn't know if I was doing the right thing. Like when I walked into the garage, I kept asking, 'Is this the one they blew up in 1993?' I'm not really familiar with the area or the buildings down there. Did I do the right thing to curl up on the tailboard? Should I have gotten inside the rig? And then I was worried about a steel girder coming down the street at 100 miles an hour, going right through the rig. All day I was witnessing things that weren't supposed to happen, so I didn't know what to expect.

"Like a building coming down in front of me. That wasn't supposed to happen. I'm trained a certain way, and I expect certain things to happen. On a typical fire, I pull up to the side of a building and there are certain things I expect to find. Okay, front fire escape, rear fire escape, one at each side, how many apartments? But everything I would normally be doing on a job was out of the question that day. I mean, we never run away from a fire. We're trained professionals. We don't do that. And here we were, literally running for our lives.

"Finally, after the second building imploded and the sun came through again, our lieutenant came back with firefighter Hoppey. He had been with another group of firefighters. It was quite a relief to know that he was okay. Quite a relief. By then we knew that the whole upper echelon of the fire department was gone, and it was complete and utter chaos.

"At some point, I started walking toward the collapsed buildings. On the way, I ran into one of the guys I work with at my regular firehouse, Engine 35. He said, 'Oh my God! You're alive!'

"He told me that they had come down in a big caravan of emergency vehicles, and as they were coming south on FDR Drive along the East River, the first building imploded. They heard the dispatcher calling the Command Center, but nobody was answering. It was complete silence. Then the dispatcher said, 'Any company at the scene, can any company contact us?' No one did, so they thought everybody was gone.

"Then they heard a guy screaming for help over the radio. He was trapped and running out of air. Because this guy was very nervous, he had a real squeaky voice, and they thought he was me. They thought I was dead. It turned out that the guy got out because, when the second building imploded, it knocked his rig over and gave him an escape route. He was saved, at least. There were so few of them who were.

"I went down with this guy from Engine 35 and met up with the rest of my company. They had stretched a hose line to Tower One. The members of our truck company were there, too, digging for survivors, and the lieutenant asked me to find a search rope. There was a rig there that was partially crushed, so I got into it and found a search rope and a flashlight.

"Three companies went in to search a building, and pretty soon they came back out. They said that they could only go in about 25 feet, and they couldn't go any farther. They'd found part of a search rope inside the building, but the other end was buried in the rubble. Who was on the other end of that rope? I found out later that this building was the parking garage I had run into when the first Tower fell.

"At that point, Seven World Trade had 12 stories of fire in it. They were afraid it was going to collapse on us, so they pulled everybody out. We couldn't do anything. There was no water pressure, so they had to bring in the fireboats. We were looking at 12 stories of fire, and there was nothing we could do to put it out.

"What we were trained to do and what we were able to do that day were two completely different things. The total sense of uselessness I felt all day, knowing that all those people were in those buildings and we couldn't get to them, that we couldn't help them, was unbelievable. And there are still so many people missing. One of the drill instructors I know lost his brother. Every spare moment he has, he goes down there and he digs. Maybe he'll never find his brother, but he's got to do something. It's a total feeling of uselessness, but at least he can say he tried.

"I was talking to a guy from Engine 53 one day, and he said to me, 'You were at the World Trade Center, weren't you? Engine 91? You came pretty close.' I said, 'Yeah, I figure 30 or 40 seconds.' He goes, 'No. It was more like two or three seconds. When you were standing at the Command Center and someone yelled 'Run,' everybody at that table was killed. Everybody who ran to the right was killed. Everybody who ran to the left or into the garage was saved. It really was just a matter of a few seconds.'

"So many really good firefighters were killed. Guys with little kids. I go to a memorial service and a guy's got three kids. The oldest is six. It's not right. Is there a divine plan? I don't know. A lot of people have flocked to churches. I'm Catholic, but I haven't been to church in a couple years, and I feel that it would be hypocritical to go now. I guess, philosophically, I've just been trying to avoid the whole thing. I can't imagine why I would have been chosen to live over anybody else."

NEW YORK STATE

MERCEDES RIVERA

AGE 23
EMERGENCY MEDICAL TECHNICIAN
ST. CLARE'S HOSPITAL AND HEALTH CENTER
MIDTOWN MANHATTAN

When we were told that we ought to speak with Mercedes Rivera about her numerous close calls with death on the morning of September 11, we never imagined that someone so young and so slight would show up at our door for the interview scheduled on Thanksgiving Day. Barely 23 and looking all of 100 pounds, Mercedes certainly didn't give a first impression of someone with such an immense and courageous spirit.

But despite her youth and diminutive stature, Mercedes demonstrated remarkable strength and wisdom in the face of constant and imminent danger. As the buildings fell and the fires raged, she returned time and time again to the heart of the disaster, determined to help people even as her colleagues urged her to leave the dangerous scene.

Mercedes spoke frankly about the overwhelming fear she experienced that day, and we were deeply moved by her honesty. If it's true that courage means taking action in the face of fear, then Mercedes is surely one of the most courageous women we will ever meet.

"When I finally spoke to my parents at the end of the day," she said, "I cried, and they cried, and my mom said, 'We don't want you working that job!' It's not the kind of job for you!' And I said, 'Well, Mom, it's what I want to do.' As traumatic as it was, as much as I saw that day and the fear I had to live with, I can't imagine having not been there. A lot of the jobs we do, we don't choose them. They choose us."

"I curled up, I waited for those windows to blow, I waited for the ambulance to turn over, I waited for death. Once again, I said my blessings. I said, 'Jesus Christ, please let my parents know I love them.'"

—Mercedes Rivera, November 22, 2001

"I work two jobs as an emergency medical technician. I work on an ambulance for Beth Israel Hospital, subcontracted by MetroCare Ambulance Service, and I work for St. Clare's Hospital and Health Center. Most of us survive on more than one job, because we don't get paid that much. On September 11, I'd just picked up an extra shift at St. Clare's after having worked a 12-hour shift the day before for Beth Israel. They'd called and said there was an opening if I wanted it. I said, 'Sure. I can use the cash.'

"My partner, Paul Bonacci, and I started our day at 8 A.M. like every other day. We called Communications, logged on, advised them of our shield numbers, and gave them the number of the bus—that's our slang for 'ambulance'—we were working.

"We did our usual routine. We got his iced tea at Dunkin' Donuts, and then went to a deli for my Spanish Latin coffee. As soon as we entered the deli, I heard one of the downtown units screaming on the radio, 'Central, I have a priority message. Be advised, I just saw a plane hit the World Trade Center.'

"My heart just stopped. I peered through the window at my partner, who looked as pale and scared as I was to hear this. I shook the radio at him and he nodded, so I grabbed my cup of coffee and ran back to the bus. The dispatcher said, 'Repeat your message,' and it was silent for a few seconds. Then he repeated the message, and all we heard was total chaos.

"I've worked the 911 system for two years, and I've been an EMT for four years, so I've worked with many, many partners. Unfortunately, a lot of them perished on that day. One of the guys I worked with, Mario Sontoro, was working for New York Hospital that day, and I heard him on the radio saying, 'Central, put me on that job.'

"My partner tuned in to the police radio, then said, 'Oh wow. It's confirmed.' My heart was pounding, not knowing what to expect. I was trying to get through to Dispatch, and then I heard, 'Everyone, just go!' The plane hit at 8:46 A.M., and this must have been a few minutes before 9 o'clock. So we started heading onto the West Side Highway.

"As I passed by St. Vincent's Hospital below 14th Street, Paul said, 'Wow. Look at that plane. It's pretty close.' I said, 'Jeez, you're right. Look at the size of that plane.' We watched that plane get closer; it flew over us, it went over the Hudson River, and then we saw it make a tilt and go right into the South Tower. It was like science fiction.

"Then we saw the explosion. We were behind another St. Clare's ambulance, and I tried to maintain control and concentrate on the road so I wouldn't smack into the bus ahead of me. We had our sirens blasting, and I don't remember hearing anything but the sirens and my own heart pounding. I knew this was big. We were trying to get an understanding of it, but there was none.

"We finally got near the Borough of Manhattan Community College, which ended up being one of the triage centers, and I had that fear in me like I did not want to go any closer. We were a few blocks away from the World Trade Center, and I was feeling panic in my heart, like I wasn't prepared for this, no matter how much training I'd done.

"We stopped and got off the bus and I saw my co-workers, Byron and John, from St. Clare's, who'd been on the bus we were following. I said, 'We can't park these ambulances any closer. We've got to turn these buses around and face north so we can get the hell out of here if we need to.' Then an officer from the FDNY said, 'No. You're too far away. You need to bring the vehicles up closer.' So we drove into Battery Park Plaza and they parked us on Vesey Street. We were right across the street from the World Trade Center.

"My partner jumped out and grabbed all our equipment, the triage tags, and our helmets—everything we could possibly throw onto a stretcher. I decided to leave the doors unlocked and the keys inside in case we had to leave quickly. I saw a New York Hospital bus with the passenger door open, so I closed their door and asked somebody, 'What happened to this crew?' They said, 'They went running in.'

"I was seeing debris falling, smoke, chaos, and fear. The fire department and the Emergency Services Unit of the police department go through regular drills to prepare for terrorist attacks, but being in the private sector, we were never trained for something of this magnitude. I went through the fire department training for my EMT refresher, because I wanted that additional training. I remember them drilling it into our heads, 'If the scene is not safe, you do not enter. You have no business putting yourself in harm's way.'

"We met up with some other EMTs and paramedics under a walkway that connects the World Trade Center to the West Side Highway, and we stood underneath this so we wouldn't get hit by debris. It was not just debris that was falling; it ended up being people, too. An FDNY-EMS captain told us, 'You guys gotta go in.' His goal was to set up triage in the lobby of the World Trade Center. At this point, we all stopped and froze and looked at each other.

"I thought, *How the hell can we enter that?* Somebody said, 'Captain, we can't go in there. It's not safe.' I guess he had his orders, and he said, 'Well, I'm giving you an order, and you have to go in.' We just could not believe it. As much as we had fear in us, as much as none of us wanted to go, we were gathered there with all our lifesaving equipment, our bright orange helmets, and our fear. Yet we also had compassion, because we wanted to help. We knew that people needed help—hundreds and thousands of people.

"So we listened to his order and made our way in. I just remember the fear, the constant sounds of things falling, and the flying paper and debris. I took the front end of the stretcher, Paul took the back, and it wasn't a smooth ride rolling the stretcher in. It was a constant carry. Me being as short as I am, I was constantly lifting the stretcher with all the equipment on it over the hoses and the debris and glass from the windows of the World Trade Center, which seemed like it was up to my knees. At one point I was trying to struggle to lift the stretcher over the hoses, and I saw a typewriter fall inches away from my partner. A typewriter in flames.

"I tried to hurry because I was so scared, not even knowing if we were going to make it into the building. We finally made our way in, and it was total chaos. Firefighters were running in, civilians were running out, water was flooding from the sprinkler system. At one point the oxygen duffel bag fell on the floor and was soaked. As we made our way in, the only patient I remember was this guy in his 60s or 70s named Bob. He was pale and diaphoretic; he'd just come down 59 flights, and he was short of breath. I stopped for him. One of the FDNY officers said, 'What are you doing?' I said, 'I can't just leave this guy here,' and he said, 'Well, we have hundreds of patients. You can't just stop for one.'

"But I couldn't just leave this one poor guy here, who looked like he was ready to have a massive heart attack, so my partner said, 'Take the stretcher and stuff. I'm going to help him.' Paul grabbed Bob's arm and helped him walk. Then I saw a burned woman in a sitting position in the lobby, as if she was still typing behind a desk. I knew there was nothing I could do for her. She was already dead.

"We all stopped at a concession stand, and we grabbed bottles of water, throwing them on our stretchers, putting them in every pocket. I said to myself, *There's no way we can set up triage in here.* I said to Paul, 'What in the world are we doing here?' We finally met with other officials from the fire department, Secret Service, and Port Authority Police Department, and everyone was trying to figure out what was going on. One of the Secret Service guys said, 'Let's just move out of here and go to Seven World Trade. We'll set up triage in the lobby over there.'

"The exit to cross over to Seven World Trade was in the middle of Vesey Street. We stalled for a second because we saw big pieces of steel falling, and then we just ran. We took our stretcher and rolled it and ran. I was constantly looking back at my partner, hoping he was okay, because he was still aiding poor Bob. They threw Bob onto one of the other stretchers and started running with him, too.

"We made it to the front of Seven World Trade, but it had already been evacuated and the doors were locked. A firefighter near us grabbed his helmet and started using it like an ax, breaking the glass. My partner used his radio, and everyone was trying to break the glass to gain entry; we wanted some sort of feeling of safety so that nothing would fall on us.

"We got in, and it was decided that this was going to be our new designated triage area. When I first started working MetroCare, we used to do a standby at this building in the evenings when people were using the gym, in case something happened. I remembered that they had a medical room in the gym with backboards, defibrillators, splints, and all this other medical equipment. I told the lieutenant about it, and he sent a bunch of firefighters up to get it.

"The lobby of Seven World Trade is like cathedral glass. I told one of the medics, 'You know, if one of those Towers goes, the impact is going to cause these windows to explode. If that happens, we're dead!' I was in constant fear, not knowing if I was going to make it anywhere alive. No matter where we went, there was that unsafe feeling. I was telling the medic this, and I told the lieutenant, and they were contemplating moving us to the auditorium on the third floor. Then a Secret Service guy said, 'There's a loading dock out back. Why don't we move out there?' I felt more comfortable with that, because at least there was a big exit. I kept thinking we needed to be someplace where we could evacuate quickly if we had to.

"All this time we'd been moving constantly, trying to get organized, trying to set up triage. We got out onto the loading dock and finally started to get the ball rolling. The loading dock was on the corner of Vesey Street and West Side, and it had good access for the ambulances to roll in, pick up patients, and roll out. We were treating Bob, doing vitals on him, and getting ready. We'd communicated our location to fire and police officials, and we were expecting hundreds of patients to come in.

"Then one of the Secret Service guys said, 'Listen. The Pentagon was just hit. I'm not sure, but I think a plane was shot down in Pennsylvania. And there's a third plane above us.'

"The fear. The fear. About five minutes after that, what did we hear? A big, thunderous, crash that sounded just like an engine. We all huddled together, not knowing where it was coming from, and we saw this big cloud that first seemed like red smoke, and then it turned black, a black cloud of debris rolling in through the gates of the loading dock. Everyone started running back into the building.

"I tried to climb up onto the next level of the loading dock, and as I put my foot forward to jump up on top, I fell. I got trampled by everybody who was trying to run back toward that one little entrance where we first came in the building. My partner stepped on me, and then grabbed me by my belt. Through all that chaos and everyone trying to run back to the front, we heard the shattering glass and screams of the people left in the lobby. Imagine what happened to them when those windows blew.

"So we turned around and tried to run back to the loading dock, but by this time we were covered in blackness and we couldn't see anything—not even with our flashlights. There was just this big, eerie, deafening silence.

"My partner grabbed my hand, and I grabbed what I assumed was John's hand, and my partner grabbed on to Byron, and Byron was leading us out by feeling his way around the trucks at the loading dock. I had my helmet on, and the visor somewhat protected my eyes, but I was still breathing in the smoke. I was trying to talk, trying to say, 'Don't let go! Don't let go of my hand!' The more I was talking, the more I was breathing this soot, and I could feel it choking me.

"At that point, I made my blessings. I'm an only child, and the most important people in my life are my parents. All I could say was, 'God, let my parents be strong for this. Let my parents be so strong for this.' I did not think I was going to make it out. I thought if it wasn't falling debris that killed me, it would be the smoke, or we were going to get buried alive in this building. It was indescribable fear.

"Finally, we saw sunlight starting to break through, and that led us out. Outside, it seemed like a big blizzard. You couldn't tell who was who. Everyone was either black, white, or gray. All the fire engines and emergency vehicles were just covered. I looked over where Two World Trade had been, and it wasn't a building on fire anymore. It was just a cloud of smoke where the building had been.

"I looked back and realized that it wasn't John's hand I was holding, it was the Secret Service guy. I said, 'Oh my God! Where's John? Where's everyone else? Are we the only ones who made it out?'

"Byron, who was John's partner, started freaking out. He was screaming, 'John! John! John! We can't leave John!' I thought, *Jesus Christ, no one made it out of the building but us.* Byron was screaming, and we were yelling for John, and the more we screamed, the more we inhaled all this crap into our lungs. I found a bottle of water I'd grabbed from the concession stand in one of my pockets, and I passed it around and we were trying to clear our eyes. Byron's eyes were so red, it looked like he was blind. He couldn't see, and he was screaming, 'We can't leave John! We can't leave John!'

"I got on my radio and said to Central, 'Seven-Nora. Seven-Nora. Be advised. We just escaped Seven World Trade. My crew is accounted for, half of Nine-Charlie's crew is accounted for, there was a Lenox Hill medic's unit that did not make it out, there were fire department EMS officials and other city officials who are not accounted for.' What did I hear? 'Seven-Nora, repeat your message.' Everything on the radio after that was chaos. It was the last time I was able to scream for any kind of help or assistance.

"At one point I walked to the corner, and this fire captain took me, shook me, rattled me, and all he kept saying was, 'Are you okay? Are you okay? Are you okay?' He must have asked me that about 20 times, shaking me so hard, and then he let go of me, walked a couple of feet, and he took somebody else and shook them up. 'Are you okay? Are you okay? Are you okay?' I saw firefighters walking by with injuries to their heads, their hands, and their extremities. Some people were carrying other people, and I kept pointing them in the direction of where the ambulances were parked.

"Now, John is about 6'3", and you can't miss this guy. In the shadows of everything, I saw him, our tall, thin John. I said, 'Oh Jesus, thank you God. Thank you.' Seeing him was such a sense of relief. We grabbed him, and he was in a state of shock. It turned out that the reason he didn't come out with us was because he'd found another exit on the loading dock, and he'd entered a boiler room. From there he found another way out of the building that led him toward the West Side Highway. He had taken off his shoes and was ready to jump into the river, thinking that was the only safe place to go.

"I asked John, 'Where'd everyone else go?' He said, 'I don't know. We had to leave Bob.'

"I said, 'Listen, we need to get into our buses and move them farther away.' As I made my way to my ambulance, I realized that the New York Hospital bus was still parked there, still running with the lights on. Later I found out that this was my old partner Mario's bus. He never made it out.

"Then I realized that all our equipment was in Seven World Trade. All I had was the helmet on my head. There was a stretcher sitting there and some bags of equipment that somebody had left behind. I grabbed the equipment, threw it onto the stretcher, and loaded it into my bus. A firefighter came up to me and said, 'Do you have some water? I can't breathe.' I said, 'Just hop in the back of my bus.' I told my partner, 'Give him some oxygen, check his lung sounds, give him some water, and let him clear the debris.'

"I drove around the block, and an EMS officer directed us to pull over. Byron and John were following us, but they were detoured and directed to park somewhere else. Looking back, I think the only reason we made it out of Seven World Trade was because we stuck together. Hand in hand, we stuck together. We did not leave until we saw John come out. But now we had to split up from John and Byron. Realizing that they'd been directed somewhere else, that they weren't behind me anymore, made me feel less comfortable, less safe. I had that ulcer kind of burning feeling in my stomach.

"People were rolling into another building now, and my partner stepped out to talk to one of the EMS officers to find out what was going on. The MERV, the fire department's Mobile Emergency Response Vehicle, was there, which was set up like a mini emergency room. I noticed another co-worker of mine who was working a different ambulance, another paramedic named John. I saw that he was getting overwhelmed by patients coming into his bus. I said, 'John, do you need a hand?'

"At that moment, we looked up the street and saw a fire department captain collapse. John looked at his ambulance and saw that his partner was working up about five patients, and he said, 'I'm going to take your bus. Let's go get this guy.' I screamed for my partner, and he jumped into the back of the bus. I jumped into the passenger seat, John jumped into the driver's seat and kicked the bus in drive, and we started rolling up toward the West Side Highway to get this fallen captain.

"We were just about to get out of the bus, when all of a sudden we heard that big, thunderous, crashing sound again. It seemed like red, purple smoke followed by black. It was like a tornado. Our bus started to shake. It was rocking back and forth. I thanked God that we were in a box ambulance that day, because there's no doubt in my mind that if we'd been in a van it would have flipped over.

"I curled up and was waiting for the windshield to crack and explode or for the bus to turn over. John had been in the military, and I remember him saying, 'This is worse than the war. This is worse than the war.' I was screaming at John, 'Reverse the bus! Raise the windows and reverse the bus!' He kicked the vehicle in reverse at like 80 miles per hour, not knowing if we were going to hit a building or people, because we were in total darkness and couldn't see a thing. He was panicking, he was raising the windows, and finally I said, 'Stop the bus!'

"Once again, I said my blessings. I said, 'Jesus Christ, please let my parents know I love them.' I was trying to contact my parents in my mind, because I didn't think I was going to make it. I waited for those windows to blow, I waited for the ambulance to turn over, I waited for death. We heard explosions everywhere, and then once again, complete silence.

"When the bus stopped shaking, I looked back and saw Paul in a fetal position on a little small side step in the rear of the bus. He was just there with his head between his legs. I said, 'Paul, are you okay?' and he said, 'I'm okay.' The firefighter who was in our vehicle earlier had left behind one of his air tanks with a one-man valve, so Paul grabbed it and was sucking on it. He passed it to the front, and we shared the oxygen, and then we shared a bottle of water, not knowing what to expect when we opened the doors.

"John turned on the headlights and the windshield wipers and cleared the debris, and we started to see, over time, shadows of people walking around us. It was still like a tornado out there. I said, 'If we get out and swallow this, it will kill us.' I said to Paul, 'Grab whatever masks we have, whatever water we have. We're going to open the doors and get people into the bus.' So we opened the doors and screamed at people to come in. We saw one person limping, and we got out and helped that person to the back of the bus. We had news reporters, fire captains, police officers, and civilians. We loaded up the back of that bus with more than a dozen people.

"John got out of the bus and met up with his partner, and I drove my bus to the MERV and delivered our patients there. I noticed that most of the injured were rescue workers, and I asked, 'What happened to all the civilians?' Then it hit me that most of them didn't make it out.

"At this point I found myself looking for my co-workers. I thought, *Where are Byron and John? Where's Tito? Where's Jerry? Where's Mario? Where's Keith? Where's Yamel?* I started to feel sick to my stomach. Some I eventually saw, and some I didn't, and I just hoped they were at some other triage sector. But that lump remained in my throat from that moment on, because I knew that if I was that close to death, not once, but twice, then my co-workers may not have made it out at all.

"It was hard to tell who was who, because everyone was covered with soot. The only way to tell the officers was because they had the white helmets. I wanted to help get things organized as units were bringing the patients in, so I asked an EMS captain, 'How are we setting up the triage?' She said, 'Don't worry about it, because we're going to be evacuating out of here.' People were running out, holding their own IV bags.

"I looked around for Paul, but I couldn't find him. I thought, *I hope he's not trying to be a hero somewhere.* I didn't want my partner to leave my sight. I didn't want anyone I knew to leave my sight. I finally found him and said, 'Listen, this isn't safe. They're going to be evacuating us.' And he said, 'Okay. Where to next?' Our ambulance was one of the first ones parked in front of that building, and they said, 'We're going to give you the first two patients.' So we ended up taking a fire captain and a civilian to the hospital.

"I was just scared. I live in this area, I know this area, but I couldn't determine what was what. I felt lost in there. I couldn't recognize anything. And now I had the responsibility of making sure these people and my partner got out safely. I was making my way out around all the detours and emergency vehicles, and as I looked out, I saw that cloud coming out of the World Trade Center. I said, 'My God. What has happened?' This is where I grew up. Since I was eight months old, I could see the World Trade Center when I came out of my building. Whenever I saw those Towers, I knew that's where home was. But now, there were no more tall skyscrapers. I never thought I'd live to see this day.

"When we got to St. Vincent's Hospital, I saw a parade of doctors, nurses, and staff waiting there with stretchers. They opened the doors, they took the patients out, and the next thing I knew, they were handing us all this equipment and water, ready for us to go back in. The question was, were we ready to go back in?

"I looked at my partner, and he looked so traumatized. He said, 'I don't want to go back.' I said, 'Paul, we have to. We're going to restock, we're going to grab equipment, and we're going to go back.' All I was thinking was, *We can't leave our co-workers behind. We can't leave behind our firefighters, and we can't leave our people behind.* It's not that we weren't concerned for the civilians, because we were. But in this field, we interact so much with fire personnel and police, we've become a family. We look out for each other. We understand that we're not always granted safety in our environment, and there's just that bond.

"I saw a priest, and I went up to him and said, 'Father, please pray for us. Pray for everyone there, pray for our brothers and sisters. Pray for our strength and guidance.' He put his hand on my shoulder, and he said a prayer. He handed me a bottle of water and said, 'God bless you. Go with God.' All I could think was, *I haven't spoken to my parents.* I gave him my parents' number because our cell phones were dead, and I said, 'Just tell my parents that I love them very much, that I'm okay, and that I'm doing my job.' Paul did the same thing.

"We headed back, and even from St. Vincent's, a mile away, we could see the debris. Eventually we were stopped at another triage center. Mind you, none of the equipment we had on the bus was ours. It was all equipment we'd found at the site that was left behind by another crew or restocked from St. Vincent's. The debris in the bus was inches thick.

"One of the St. Clare's Hospital Jeeps arrived, and out of this Jeep came about 10 of my co-workers. Everyone was lifting us up and giving us hugs, and they were crying to see Paul and me, because the last they'd heard of us, we were in the World Trade Center. That feeling I had that day, to feel the love and emotions, to know that they were okay, was unbelievable. They said, 'You're going to be okay. You guys are strong, we're going to stick together, we're not going to leave each other's sight.' That was so reassuring to hear. I felt like a lost child who'd just reunited with my parents.

"Then a lieutenant came and started organizing how we were going to set up triage in the street, and the next thing I knew, a police officer came up to us and said, 'Does anyone own a black briefcase?' We said 'No,' and he went on his radio and called the Emergency Services Unit. The bomb squad showed up, and they detected a possible bomb. They had us running back to our vehicles, and we moved about three blocks up. I said to Paul, 'When is this going to end? When is this constant fear going to end?' We looked at each other and he said, 'What are the chances we're going to make it out of here alive?'

"So we all finally met up again, more ambulances were coming, more people were reuniting, and we were pulling out all the equipment we had in our buses. There was no officer there yet to set up the triage area, so the whole group of us put our heads together and started to run operations ourselves. Then somebody said, 'We might have to relocate again, because there's a gas leak over on Chambers Street.' I said, 'How many times are we going to have to evacuate?'

"At that point, they said that Seven World Trade had no face and it was ready to collapse. I remembered the firefighters I had sent upstairs to get equipment. The medical office where I sent them had windows that faced the World Trade

Center. The guilt started to creep up on me, and I thought, *Did those firefighters make it out alive? What about the other people we left behind in that building?*

"We had escaped death three or four times that day. We were covered in ash. The only thing we had that day were short-sleeve shirts, our stethoscopes around our necks, and these bright orange helmets. We still had soot around our lips and our eyes, even after we'd washed them out several times. One of the guys who works with us told the lieutenant, 'This crew needs to go out of service.'

"We transported a deaf woman who had a sprained ankle back to St. Clare's, and when we got there, she hugged us. I guess it was obvious by looking at us, the expressions on our faces, our body language, the soot on our clothes, and the mess in the ambulance, that we needed those hugs. Our coffees from the morning remained in the holders on the side doors, full of debris and ash. I couldn't believe what I was leaving behind.

"As we were driving up West Side Highway, there was one person holding this big sign that said, 'Thank you EMS.' I think the fire department EMS got more recognition just because of the logo they wear that says 'FDNY,' as opposed to us, where we're private. Maybe we get a few thank yous in the store, but if I walk in with a co-worker with a FDNY jacket on, all the praise and glorification goes to that person. And it hurts. It really hurts. It doesn't make the healing process any easier.

"We lost six co-workers from the private companies and two from FDNY-EMS. Being in the private sector, we aren't considered municipal employees, and we aren't eligible for the same funds received by people employed by the FDNY. The actions of my co-workers made them heroes in my eyes. They left behind families and kids and financial problems. It makes their deaths that much harder.

"I lost Mario Sontoro, and I think his death is the one that hit me the hardest, because we worked with each other so many times. His daughter just turned two. When you work with someone 8, 12, or 16 hours a day, it's amazing the bond you create with your partner. We go to jobs not knowing what's going to happen. How many jobs do we get called 'other,' where we have no information? Maybe it's a drug deal that's gone bad and we don't know that. We're rolling up with no police there, no bulletproof vests, with just a bag, some lifesaving equipment, and the compassion in our hearts. How many times have we run into a burning building before fire and police get there?

"My co-workers who died weren't heroes just that day, but every day they came to work and did their jobs. How many times have we put ourselves under a vehicle, trying to save a child who's been pinned underneath, not knowing if that vehicle could come down on us, too? How many times have we climbed up on a fire escape because we think there may be a child left alone in an apartment or an elderly person who's fallen? How many times have we dashed to help an injured cop, not knowing what we were walking into?

"I turn on the TV, and I'm waiting to hear something about EMS. I'm waiting, and I'm waiting, and the hurt, the pain deepens. We put ourselves in the line of fire just as much as the cops and firefighters, and I'd like people to give my co-workers that same respect. The days pass, and the weeks pass, and nothing. God knows, I give the firefighters credit and all due respect to the police officers. But there were other people there, too. After a while, the anger started to build up and accumulate, and I started to find myself disgruntled for the first time.

"Several days after the collapse, Paul and I went down to Ground Zero. There were about 600 volunteer EMS workers from around the world waiting to go in and dig, but they wouldn't let any of us in. That made me mad. I was good enough to send in when the building was on fire. I was good enough to send in when people were trapped. But now that I wanted to be there to help look for my co-workers, I couldn't go in because I didn't have an FDNY patch on my shoulder? They weren't letting anyone in unless they worked for the fire department, and to this day, that's how it is.

"I was doing pretty well until two days ago, when I went to Mario's memorial service; then it hit me all over again. It was a good feeling, though, to see all of us in uniform, to see the fire department EMS there in uniform. It doesn't matter whether you work for the FDNY or you work Lenox Hill or you work MetroCare. It shouldn't be about that. We all are here to do the same job for the same reasons. Whether you were my partner or not that day, I would have grabbed your hand, and we would have made it out together.

"One of the last times I worked with Mario, I was upset because I'd lost my driver's license, and he said to me, 'Sweetheart, never lose hope. You can't lose hope.' Those words lingered. When they said that Mario didn't come home, it was like a replay in my mind of Mario saying, 'You can't lose hope.' I wanted to believe he was stuck in a sublevel somewhere, eating McDonald's, helping people, making people laugh. I wanted to believe he was keeping them calm, saying, 'Sweetheart, never lose hope.'"

JoANN SPREEN

AGE 31
POLICE OFFICER
NEW YORK CITY POLICE DEPARTMENT
1ST PRECINCT SCOOTER TASK FORCE

Our interview with JoAnn Spreen took place at Police Headquarters, where she arrived in street clothes on her way to a doctor's appointment. Still off-duty as a result of head injuries she received on September 11, JoAnn spoke quietly about her ordeal and often touched her fingers lightly to the healing gash on her forehead. It was obvious to us throughout the conversation that she was very aware of the physical reminders of what had happened that day.

Born and raised on Staten Island, JoAnn is the youngest in a large family and maintains close relationships with her family members. "It helps a lot to have people close to me who know what happened," she said. "It helps to talk to them and let them know how I feel."

Having lived in New York City all her life, JoAnn always considered the Twin Towers a symbol of her home. "I still stare with disbelief at that empty space," she said. "Even driving in today, it didn't feel like New York City. Not to me. Not yet."

JoAnn has been a police officer in the 1st Precinct for eight years, and she was back on patrol when we returned a few months after our interview to photograph her on her scooter. When she removed her helmet, however, the scar on her forehead remained visible.

"It was mostly airplane parts that were on the ground. Big airplane parts. That must have been what hit me."

–JoAnn Spreen, November 23, 2001

"I usually spend my days on a two-wheel scooter, riding around on a community policing beat in Manhattan South. Even though we're assigned to the 1st Precinct, any time there's a demonstration, march, or rally, we handle that. I'm the only woman on day tours, and there's one woman at night.

"We make a lot of peddler arrests in our precinct, and we confiscate the property. Once a week, we have to bring the property out to Queens, because we can't keep it in the precinct forever. It would overrun the place. On the morning of September 11, I volunteered to go with another guy in the truck to drop off the property. We were running behind schedule, so we decided to get something to eat quick and then go. We wound up at Chambers and West Broadway, four blocks north of the World Trade Center, when the first plane hit.

"My partner was driving, and he was about to make a left turn when we heard this noise, this crash, but it didn't sound like a car accident. It sounded like—of course—something I'd never heard before. At the time, I thought one of those big 18-wheelers had lost control and hit a building. That's how it sounded. It was tremendous. We went one block up and saw all these people looking up, but we still couldn't see anything because there was a building blocking our view of the World Trade Center. We got to the corner of West Broadway, one block north of Chambers Street, and that's when I saw the North Tower. That whole floor where the plane hit and everything above it was in flames. The whole thing was burning.

"There's a heliport nearby, and I was thinking that maybe one of the tour helicopters went off course. I don't know why I was thinking that considering the size of the hole and the size of a helicopter, but I couldn't imagine anything else.

"I turned to my partner, and we were jumping up and down in the truck. I was saying, 'Hurry up! Get down there!' It was a big truck, so it took a few minutes. The radio patrol cars were coming up beside us and passing us because they're faster. It felt like it took forever to get there. We turned onto Vesey Street on the north side of the World Trade Center complex, went one block up to Church Street, and jumped out of the truck. I was in civilian clothes at the time, so I made sure I got my shield out and we ran down Vesey Street toward the train entrance. People were just pouring out of that exit because it was also one of the exits from the Tower that was hit. But as they were coming out, they were in such shock, I guess, that they just stood there, staring up at the fire.

"We knew what we had to do. We saw firefighters running up the stairs, we saw the Emergency Services Unit going into the buildings, and we saw ambulances coming in. We knew that someone had to keep those people moving away from the scene. They were just standing there, staring up at the building, and things were falling 20 yards away.

"My partner and I went rushing up to them. 'Come on! Hurry up! Let's go!' A couple of people I had to actually push, like 'Wake up!' They just had their eyes wide open, staring. Then two other uniformed officers showed up, and we made a line so that as people were coming out of the World Trade Center, we were directing them to the next cop, to the next cop, to the next cop, shoving them off Vesey Street and north on Church. People kept stopping, and I had to keep running up to them. 'Come on! Keep going! Keep going!' By then, there were fire engines and ambulances and cop cars and everything flying down there. I kept telling people, 'Be careful! Don't hurt yourself, be careful! Just keep going north on Church Street.' That was going on for about 15 minutes.

"Then I heard the noise. I heard that noise an airplane makes when the flaps are going down. It almost sounded like a takeoff noise. And then I heard the same noise I heard earlier, that crash that sounds like a truck hitting a building. I was standing next to another police officer, Mike Gerbasi. We both looked up at the same time, and that's when I saw this orange ball of fire coming out directly over our heads.

"Right after that, it turned black. Orange flames with smoke were coming out of the building, and in the smoke you could see something shiny, like confetti and paper, like the ticker-tape parade they have for the Yankees. That's what it looked like. I don't know how long I stood there, and then I turned and punched Mike in the arm and said, 'Run, Mike! Run!'

"We started running up Vesey Street toward Church, and I don't know how far I got. I must have gotten hit right away. I remember seeing a guy trying to push himself under a car, like trying to slide under it. He was lying parallel to the car, but the curb was too high and the car was too close, so he was just trying to wedge himself in there. That's the last thing I really remember seeing. Something hit me in the head, and I got knocked down. The doctors later said that if I didn't remember, then I got knocked out.

"I had been on the north side of the building, and the second plane came from the south and went into the south side of the South Tower. When it went in, everything shot out from the north side of the building. I spoke to an Emergency Services officer who was down there on that corner that day, and he said it was mostly airplane parts that were on the ground at the time. Big airplane parts. That must have been what hit me. Things were bouncing off buildings, and Mike, the officer with me, almost lost his arm.

"People have told me that I got up and that I was stumbling. I do remember that I couldn't see, that I was bleeding. Blood was in my eyes, and I could taste it. It was just running down my face. I had very deep lacerations. An officer grabbed me and pulled me up the block. I didn't know that stuff was still falling at the time because I was holding my head and my eyes were covered. People kept saying, 'Does anybody know her?' My shield was around my neck and it slid out. Someone saw it and said, 'She's a cop.' Then this officer I work with, Louis Chiaccheri, turned and looked and saw it was me and said, 'I know her.' He and the others half dragged, half carried me to a police car.

"But it wouldn't start. There was stuff hitting the roof and the engine wouldn't turn, so they pulled me out and put me in a fire department car. They laid me in the back seat, and Louis was leaning over my head. He says that I kept telling him, 'Watch out, Louis; they're jumping.'

"Someone had told me earlier that people were jumping out of the Towers, to put it over the radio. I tried, but the radio was completely dead. It was chaos. I couldn't even transmit. There were people jumping out of the North Tower, right above where the fire was and right below where the fire was, and I guess that stayed in my head.

"Louis kept saying, 'It's okay, it's okay.' They took me to NYU Downtown Hospital, which is near the Brooklyn Bridge. They got me onto a gurney, and I kept telling them, 'There's a radio in my back pocket.' But it turned out that the radio had snapped in half. Just the battery was in my pocket, and it had cut my back open.

"The worst thing was, because I had injuries to my face and head, they covered my face with a drape so that they could clean the wound and do what they had to do. I couldn't see anything, but I could hear what was going on with the people around me. These were the first people who got hurt. These were people who got hurt from the plane hitting the building, not from the collapse. All I remember was hearing people screaming. The doctors were saying that the guy next to me had burns over 40 percent of his body. I think that's when I started going into shock.

"I was lying there, I couldn't see anything, and I was just checking everything out. I said to myself, *Okay, everything works. I'm not missing any limbs. I'm okay. I'm alive.* I was also thinking, *This hurts.* I've had accidents before because I ride a two-wheel scooter. I've been hit by two cars already, but it was nothing as serious as this.

"It took the doctor almost two hours to do the sutures. One of the cuts was down to the bone. The doctor was explaining to me that he had to sew inside and out. During that time, the buildings collapsed. When I got up from the table, there were smoke and dust all over the inside of the hospital because it's so close to the World Trade Center. It was like something out of *M*A*S*H.* It was unbelievable. It was like there was a movie going on. Not a good movie. And I kept thinking, *When is this going to stop?*

"I couldn't believe this had happened. This is my precinct. I've worked here for eight years. I was worried about Mike, the officer who had been standing with me when the second plane hit. I kept asking about him, but nobody could tell me where he was. I found out later that after he got hit, he held his arm and kept running. He was taken to another hospital, and he was all right.

"A lot of rumors were flying around the hospital. I thought of all my friends I work with and the unit I work in. Most of the guys have been there for years. It's the type of unit you get into and you pretty much stay until you get promoted or you retire. It's a good place to work. We're all very close. It's like having 30 brothers because we're all close in age.

"By some miracle, no one else I work with got hurt. They were right there when the buildings went down. The precinct is three blocks away. If the first plane had hit an hour later, I can't imagine how many more people would have been inside those buildings. It would have been 10 times worse. Between the morning rush-hour times from 9 A.M. to 9:45 A.M., you can't even drive through there. It's like a train of people. By 10 o'clock, they're all up inside the Towers.

"I was thinking about all this while I was in the hospital, and I was thinking about my father, who was very sick with pancreatic cancer. One minute I was feeling like, *Thank God, I made it,* but then the next minute I was thinking about all the people who were lost. I was mad, too. When I found out that somebody had deliberately done this, I wanted payback. I'm going to have to live with this for the rest of my life. I'm going to have to remember this. A bunch of things were going through my head.

"My sister is a lieutenant with the Pattern Identification Unit in Manhattan South. She eventually found out where I was, and she came into the city and picked me up. When she walked up, I was standing there with a bottle of Gatorade, a hospital gown on, my clothes a mess, my head taped up, and I was totally out of it because of the medicine they gave me. She got me out of there and took me home to my family.

"The first couple days I was home, they had me on so much medicine I just slept. When I started coming off the medicine, I didn't want to talk about it much. My father was an Emergency Services cop, so he's seen things. He's never really told me a lot of stories, but enough so that I knew the things I told him weren't going to scare him. So I talked to him about it. I told him about everything that happened that day.

"My oldest sister is the mothering type. I told my father about the things I saw, but I talked with my sister on the emotional level. I felt a lot better after I talked to them, because I had to talk to somebody. It was a horrible thing to see, to live through, and I was thankful that I got to make it. I got to get out of there. There are so many people who didn't. That's one of the things that was bothering me.

"I was completely off work for about a month. Fully out. Then I came back to limited, light duty, but I had to take more time off because my father was sick and I was taking care of him. He passed away last week. It was almost as if what happened, with me getting hurt, gave me time to be with him that whole month. Every day. It was a horrible thing to watch, but I really got time to be with him, to just spend the time with him.

"I haven't been officially back to work yet, but I want to get back. I still get dizzy spells, and when I look up, the room spins. People say, 'Why don't you go somewhere else?' But I don't want to go anywhere else. I want to be here. I think I have to go back, for myself.

"A lot of people have said to me, 'I can't believe you still like doing this work.' I can't help it. I have great friends to work with, and I enjoy all the people. I get to do different things every day. It's not like I have to come to an office and sit at a desk, type at a computer, and answer phones. I can do that some days, and some days I can be out on the street. I've seen things I never would have seen before. Good things and bad things. I'd like to say that my father had some influence on my career choice, he being a retired police officer. My sister's a lieutenant with 17 years on the job. My niece's husband is a cop in the 1st Precinct. So I guess the idea just passed down through the family. I like it a lot. I have a good time, and I always enjoy it.

"Three days after my injury, I had to go see the surgeon. One of the guys I work with came and picked me up, and he took me out to get something to eat. We went into this diner down the block from the surgeon's office and a guy in the restaurant came up and hugged me. That was a very odd feeling. I handle a lot of demonstrations, so I'm used to negativity. I'm very used to people telling me to go away. Of course, it won't last, but it's a very weird feeling to have people come up to me in a positive way.

"I don't worry about the little things like I used to. I think that's changed for me. The little things can take care of themselves. I used to worry about schedules at work because I'm also the vehicle coordinator. I'm the one who maintains all the vehicles. I was always worried about whether or not everything was up and running. Guys would come up to me and say, 'This isn't working,' and I'd say, "All right, I'll get right on it; I'll fix that right now.'

"But now, I'll get to it when I get to it. I'm not going to go crazy worrying about things like that. I'll worry more about personal things. Between what happened to me and then with my father being sick, everybody's pulled tighter. I feel like my family and my friends are more important than anything else, and I'll do whatever I have to do to take care of them.

"What I'm left with physically is a severe concussion, and sometimes concussions take a long time to heal. I have headaches and sometimes that vertigo feeling. I had a CT scan, and it came back negative, which is good. There's a nerve that got cut in my forehead, so I can't feel it when I touch it. It's a numb spot, like when you go to the dentist and get Novocain. That's the feeling.

"I don't want a scar on my head, but if this is what I have, I can deal with it. It's nothing, nothing, nothing at all compared to what it could have been. But every day, it's like a reminder. It's healing. It itches. It hurts. I still can't wear a hat. I don't mind the scars on my head, but these on my face … I don't want to have to look at these scars every day and remember what happened for the rest of my life."

REGINA WILSON
FDNY

REGINA WILSON

AGE 32
FIREFIGHTER
FIRE DEPARTMENT OF NEW YORK
ENGINE 219
PARK SLOPE, BROOKLYN

When Regina Wilson arrived for her interview on a rainy day in Midtown Manhattan, she greeted us with a soft-spoken introduction and a handshake that nearly cracked the bones in our hands. Throughout our early morning conversation, Regina continued to reveal these two sides of herself. She was both gentle and sweet, strong and tough.

Regina wept openly when she told us about the members of her firehouse who were killed on September 11. Compounding her loss is the knowledge that on that morning, prior to the first plane crashing into the World Trade Center, she had traded assignments with a fellow firefighter who later died in the collapse.

"I know my guys," she said of her company at Engine 219. "I know they worry about me, but I don't want to give them anything to worry about. I know they're all hurting, just like I am."

Regina recently returned to Engine 219 after completing rotations at Ladder 16 in Manhattan and Engine 289 in Queens as part of her training as a probationary firefighter.

"I was really looking forward to coming back here after I got off rotation," she said. "These guys were good to me. They were very good to me, and now not all of them are here anymore."

"I think the essence of love is laying down your life for others. If you've ever felt that nobody loves you, just look at a fire truck going by. Then you'll know that somebody does."

—Regina Wilson, January 8, 2002

"I've been with the Fire Department of New York for three years, and I guess you could say that my journey started when I filled out the application. I was working in the treasury department for a gas company, when a woman from the FDNY recruited me at a Black Expo. She spoke to me about the job, the benefits, the good sides, the bad sides, and it was intriguing to me. I gave her my name, and an application came to my house. After I returned the application, I got a card in the mail telling me where to go to take the written exam.

"The school where the exam was being given was in Lower Manhattan, so I got off the train and walked around the block. I was excited. When I got to the corner where the school was, there were nothing but white men all down the block. I was like, *Okay, this is the school, this is the place.* It was so intimidating. At first I thought there were no other women there, but then I started to see a few. You could count them on one hand.

"So I stood in line, and I told myself, *I'm not turning around. I'm not going home.* The line started moving, and we went into our designated classrooms. There was just one other woman and me in the classroom, taking our test with all those men. As soon as I sat down in the chair, that's when I knew that this was what I wanted to do. It was everything the recruiter had told me. She said it's a job that's not filled with a lot of women. It's a job that's very exciting, and I could definitely make a difference. I thought, *Yeah. I'm going to make a difference. I'm going to be a woman in the fire department.*

"I trained for the physical with a lot of women, but when I went for my 13-week school at the Fire Academy, I was the only woman in a class of 283. It was a little intimidating. I was coming from Corporate America, and they said to come in business attire, so on the first day I wore a suit and heels. That was business attire to me. We were seated alphabetically, and because my last name is Wilson, I was sitting all the way in the back, up in the balcony. The drill instructors had to know I was there, but they didn't really see me. Then we all had to go outside and line up, and they were like, 'Hurry up! Hurry up!' I was running in my heels, and the training officer said, 'Wilson! Stop running or you'll break your neck!' So right there, I had a rude awakening. Since then, those who have seen me work know me to be a strong woman, a good worker, and a good firefighter.

"I had put in for overtime the night before September 11, and I got hired for overtime that morning. Our firehouse has an engine company and a truck company. The engine has the hoses on the back, and once the fire has been located, or if we have a visual on the fire, we hook up, supply water, and put the fire out. The truck has the ladder on top. It can either be a tower ladder that some people call a 'cherry picker' because you can stand on the ladder and it can be elevated, or it can have an aerial ladder that rises up to the side of the building. The truck company goes out and searches for fire and searches for life. They force open doors and make rescues—that sort of thing. Their job is to find the fire so that the engine company can put it out.

"Earlier that morning, we had decided who was going to go to the truck and who was going to go to the engine. I was supposed to work on the truck that day, but because I was on overtime and had been on the engine the day before, one of the guys said, 'Why don't you just stay with the engine? I'll work on the truck.' So we wound up switching that morning.

"I was sitting with another guy in house watch—the area where all the calls come in. We turned the TV on, and I was like, 'What's this?' My first impression was that it was a movie, until I realized that we had turned on the news. I said, 'That's the Twin Towers!' So we sat there watching it, and then I walked to the kitchen and they were watching it in there, and somebody said, 'Wow. The Manhattan company's going to be busy today.'

"Maybe 10 minutes later, the truck company got dispatched out on the fifth alarm. They were the first unit from the outside Boroughs to get sent out because our firehouse is the closest. We're maybe about 10 minutes away from the Brooklyn Bridge and the Brooklyn-Battery Tunnel. Some of the other companies in the neighborhood got sent out, too, but they didn't send our engine. We were sitting there, still trying to figure out what was going on, what this was all about, when the second plane hit.

"I had taken over the house watch, and when I heard the guys in the kitchen yelling, I ran in there to find out what was going on. They said, 'The other building just exploded!' Then people were starting to say that they saw another plane go into the building, that it was a terrorist attack. At that point, we were just trying to figure out what we were going to do. The boss was like, 'Listen, we've got to get ourselves together, and we've got to come up with a plan.'

"I didn't know what to think. I didn't know what was going on. I felt confused, but I didn't feel afraid. I think the biggest emotion I was feeling was like I needed to help, like I needed to do something. We kept saying, 'Why are they keeping us here?' I was getting upset. The guys from our truck company were out there, and they didn't know it was a terrorist attack. They may have known there was a second explosion, but they didn't know it was a terrorist attack.

"I went back to the house watch, and I was watching the news on television when an alarm came in. It was a fire in a building on Hanson Place, the tallest building in Brooklyn. We're like, 'Aw, man, they hit in Brooklyn! They hit in Brooklyn!' We headed out, expecting the building to be on fire, because we thought they had flown a plane into that building, too.

"But it wound up being a false alarm. We got back on the rig and started going back to the firehouse, when we got another run. This time it was for food on the stove. It was very frustrating. We had to go all the way to the other side of town for food on the stove. Somebody burned breakfast.

"As we were coming back to the firehouse down Eastern Parkway, there's a point where you have a slight view of the World Trade Center. We saw a large amount of smoke and people standing on the street. It felt like the world was coming to an end. Everybody was in a stupor, people were pointing, and it felt so much like a movie. I felt like I was outside myself, like not a part of it, but watching it and not really knowing what to do.

"When we got back to the firehouse, we were dispatched to Manhattan. We had a covering officer who wasn't our regular officer, but he was very good. He was telling us, 'Stay together, keep your heads, make sure you listen to me, and I'll make sure that everybody's looking out for each other.'

"We were driving through the tunnel, and I was thinking, *I'm not going to make it. I'm not going to get out of here alive.* I said, 'God, just get me through this tunnel, because if they're blowing up the bridges and tunnels, we're messed up.' All I kept thinking was that they were going to blow up the tunnel, and water was going to come gushing in. So I was like, 'Lord, just get us out of this tunnel.'

"There were no cars in the tunnel, because the cops had gotten there and stopped traffic, so we just breezed through. But when we got to the end of it, there was this big explosion, and we thought they blew up the front of the tunnel. That was when the first building fell. A white cloud came and pushed through the tunnel. It was pushing right at us, and I remember the rig shaking. We rolled the windows up, because we didn't know what it was. All I kept thinking was that they blew up the front of the tunnel, and we were going to have to dig through rocks and move boulders to get out.

"As the cloud started to dissipate, we could see light, so I knew we weren't trapped. The officer said, 'Everybody get off the rig, grab your gear, stay close, and wait for me.' We left our engine in the tunnel, because we wanted to make sure there was a safe place to bring it out. The engine is our lifeline. We have 500 gallons of water on there, so if anything happens, at least we have that reserve until we can hook up to a hydrant.

"The boss said, 'Listen everybody, reserve your supply of air. Don't put your mask on. Take your hood and put it over your face, because we don't know how long we're gonna be out here. We don't know if we'll have to go into a building and save somebody.' So we put our hoods over our faces and our hands on each other's shoulders and walked out in a straight line until we started to see more sunlight and it started to get a little bit clearer. We left the chauffeur on the engine and walked out, carrying our emergency medical supplies with us.

"Once we got out, it looked like a ghost town. It looked like an abandoned city. We started to check cars, because we wanted to see if people were in them, if anybody was hurt. When we found people in cars, we told them to sit still until we could figure out what was going on. Then all these people started to walk toward us, and they were covered with this white powder. We saw a woman who had asthma, and she could hardly walk. There was nothing we could do for them but to send them down the tunnel so that they could get out of Manhattan and go to Brooklyn. They looked like lifeless zombies. They were walking so slowly, and they looked like they were wrapped up in cloths, like zombies are wrapped up in the movies. They were in a daze, so confused. And I couldn't even help them. I couldn't safely take them where they needed to go. At that point, I had the first feeling of helplessness. I knew I had to stay there, and I couldn't make sure that nothing would happen to them in the tunnel. I just wanted to make it better for them, and I felt so bad not being able to help them.

"By then, our chauffeur had seen other engines veering off to the left, so he pulled the rig out of the tunnel. The officer decided that we were going to find the Command Post and get our assignment. We were a few blocks away from the World Trade Center, but we couldn't tell exactly where the streets were. We saw a lot of engine companies and truck companies working in one area, so we started to walk in that direction. I had the control position, which means that I open up the hydrants, make sure we have water supply, and hook up the hydrant to the engine. I also carry a radio to communicate with the officer.

"I had my radio on, and I could hear all the different maydays of people trapped. I could hear urgents where the supply of water had been cut off and nobody could get any hydrant water. Then I heard, 'We're under attack! We're under attack! Everybody run!'

"We couldn't really see, because it was like a big, dense fog had fallen. The boss said, 'Everybody get back to the engine!' We started to run back to the engine, and then we heard this sound. It sounded like a plane, and it seemed like it was hovering over us, and the next thing we heard was this big boom. It was actually the second Tower falling. We turned around and saw a big black smoke cloud with flames in it coming right toward us. The boss was screaming, 'Get behind the rig! Get behind the rig!' I remember the chauffeur calling out to us, 'It's coming! It's coming! The smoke is coming!' And the boss was like, 'Hurry up! Hurry up! Everybody put on your mask!' and making sure everyone was okay.

"The next thing we knew, this black smoke came and pushed against us and covered us, and everything went completely black. Four of the five of us were standing in back of the engine. The boss was talking to us, saying, 'Everybody just stay calm, stay calm. We're going to be okay.' It was so dark, I couldn't see my hands in front of my face. I couldn't even see the lights on the rig, it was so black.

"I thought I was going to die. That was it. I was thinking about my family, and I was praying. We were all praying. Even the guys were saying, 'Everybody, let's just stop and pray. I don't know if you believe in God or you don't, but we need to stop and pray.' And that's what we did.

"I think that's when the realization of what my job really is hit me. Up to that point, I went to fires. I even got injured at a job. I knew that the job was dangerous and that I could die, but that's something I had to put in the back of my head in order to get up and do my job every day. I had gotten very good at doing that. I love my job, and it didn't really occur to me how dangerous it was. But at that point, I was trying to figure out what my family was going to do. I thought I was never going to see my family again.

"I owe so much to that covering officer, because none of us knew exactly what to do. But he kept his head. The chauffeur did, too. They kept their heads, and I felt so comfortable with them. I felt like, if I was going to die, I was going to die with them, and I was content.

"Finally it started to clear up, and the boss started asking, 'Is everybody okay?' At first, we thought one of our guys was missing, but he wound up being inside the engine.

"I was trying to sum everything up. I was trying to comprehend it, but I couldn't. You wouldn't think this was New York City. This is such a vibrant, energetic city, but it looked lifeless. It looked like one of those futuristic movies where they show the way-off future, and the place is desolate and barren, and people are craving for water. It was horrible. There were shoes and papers everywhere.

"We grabbed all our tools and extra bottles of air, because we didn't know if we'd have to do any saves. We didn't know if we'd have to force doors, so we brought all the tools we could carry. We started walking, and I heard all these tires exploding, saw all these cars on fire, buildings on fire, and it felt like a war zone. On the radio, I was hearing all kinds of maydays. I was hearing them asking for body bags. I was hearing people screaming, 'Help me! Help me! Get me out!' Then someone would ask for their location, and they'd say, 'I don't know where I am! Just come and find me!'

"There were fires going everywhere. In one of the buildings, there was fire on every floor. They needed to find a tower ladder to fight the fire in that building, but there were no tower ladders in the area at that time because most of the trucks and engines had been destroyed when the buildings fell on them. A lot of the rigs coming in were fresh engines, but there were no tower ladders. So they wound up trying to fight the fire from inside the hotel next door. They ran lines through the building and were shooting water out the windows at the fire. It took a lot of time to get all that set up, because they were trying to find hydrant water and all the water was cut off.

"Rigs were everywhere, and we couldn't get our engine close enough to where we were. There were several abandoned rigs in the area, and we had to take another rig we found that was covered with debris. We couldn't use certain parts of the rig because it had large chunks of concrete on it and there were crushed cars pushed up against it, but it was basically operational.

"There was a water tank in one of the buildings, so we hooked up to the standpipe and got water that way. We used the deck gun, which is a nozzle that sits on top of the engine and moves so that you can shoot water greater distances. It has a height to it, so instead of you being on the ground with the nozzle, it's fixed on the top of the rig and can reach higher places. It also has a larger degree of pressure, which means it can go a little bit faster and has more of a water flow on it. We were the only ones having water flow in that area, and it wasn't even off our own rig. We were the only operational unit in the area.

"Eventually they found a tower ladder, and we needed to get a larger supply of water to attack the fire in that building. I couldn't find a workable hydrant, and we couldn't get any more water, so they called in the marine unit. We had to find a 3½-inch hose and run it from where we were to the river. I would say we needed about 10 lengths of hose, and a hose is about 50 feet long. So we walked around and found abandoned rigs and took the hoses off those rigs. We connected them down to the river so that the marine unit could draft water from the river to supply our rig and the tower ladder.

"I knew we had to put these fires out. All I kept thinking was, *It's going to be worse if we don't put these fires out, because there could be people trapped in these buildings. There could be people trapped in these cars.* Everybody got into work mode, trying to make sure we had hose connections, trying to find hydrants. I don't think it had occurred to us yet that the firefighters who'd come in on these abandoned rigs might be dead and that they weren't coming back.

"About that time, a group of our off-duty firefighters found us. They'd gotten up early that morning for a golf outing, but when they heard that the World Trade Center was on fire, they jumped in their cars and came back to New York. They saw that their city was in trouble, and they flew back at 90 miles an hour. It was amazing that they found us, because that area is so huge. I was so happy to see them.

"Another blessing was that I saw Adrian Walsh, one of our female firefighters. It was so good to see another woman's face. I felt a sense of comfort then, because she's somebody I can relate to. She was trying to find her guys, and she'd gotten the news about our truck company. She told us that our truck had been destroyed and our guys were missing. At that point, I knew they were dead.

"The chief assigned another company to take care of the engine, because we had been there most of the day. The captain of our house gathered us all together and was trying to comfort us. He was trying to figure out what everybody was thinking, what we'd heard. We wound up going down to the pier to regroup, because it was a lot less smoky there. We were all sitting together, trying to figure out what we were going to do next. We were getting our eyes flushed, we were coughing, and trying to get our bearings. We'd been running around, being so busy, that we hadn't had time to take care of ourselves. But at that point, there was nothing more we could do. We had to stop and let them rinse out our eyes and just kind of get ourselves together.

"Then we got another assignment. We were sent back up the hill to the hotel where they were fighting the fire in the building next door from the windows, and we were supposed to take over the hose line to relieve that company. They were on the thirty-second floor, and we had to walk up the stairs because the elevators weren't working. It was something like 10:30 at night. By the time we got up to the twelfth floor, the chief had given the order that everybody who had been working the day tour was relieved, so we left.

"Somebody wound up taking our rig, and they found a hydrant that worked, so they hooked up and used our rig to supply the hotel. We had to leave it there. I remember us trying to find a ride back, and we ended getting on this big yellow school bus. It was quiet. It was very quiet on that bus. People were talking, but they weren't really saying too much.

"There were so many people from the neighborhood in front of the firehouse when we got back, all the little kids and the neighbors. They were looking to see who came back and who didn't, trying to make sure everybody was okay. I went in and saw some of the guys, and their eyes were so puffy from crying, because they had just found out that the truck was gone. We were the first people to come back, and we were covered in dirt. I remember taking off my gear and putting it down, and one of the guys hugged me, and I just felt limp.

"Out of my house, we lost a total of seven firefighters. Every day I think about John Chipura, who traded rigs with me that morning. I think about how it could have been me, and I don't know if it should have been me. Every day I think about how he must have felt, what was he going through, being crushed like that. I have to see what his family is going through, and I think about my family having to go through that.

"I hear people saying, 'You can't have feelings of guilt,' but it's hard not to. When I first came to this firehouse, John was the person who took me under his wing. He was a probie—a probationary firefighter—just like me, but he was so helpful and so warm. He cared for everybody, and he always tried to make sure everybody was fine. I looked forward to working with him, because he was always so good to me.

"This can be a cruel job sometimes. Some people can be very heartless and say some very mean things, unfairly, unjustifiably. And then you have people who will always have your back. I remember when I got burned on my knee and was in the hospital, Henry Miller from our firehouse would call every other day. He called me constantly just to see how I was doing. And now he's gone.

"And Frank Palombo, I remember just a few days before this happened, he called me in and said, 'I know it must be hard for you sometimes, being the only woman here and being a minority. I know how some of the guys can kid.' He was asking me if I was okay, if I needed anything, just trying to talk to me, trying to figure out where my head was. And now he's not here anymore.

"Do I think about quitting? No. Never. There's still too much to do. I love my job. I love the work. It's one of the best decisions I've ever made. I love the camaraderie, and I wouldn't trade it for the world. I think that's why I went back and took the detail at the World Trade Center. I went down there for 30 days in December, because I said, 'I have to find Henry. I've got to find Frank.' I didn't know what else to do. I felt useless being at the firehouse.

"I want people to know that the men and women of the fire department are people who, every day, make a conscious decision to risk their own lives to save other people's lives. Every day, they wake up and leave their families, leave their wives and husbands, leave their children, leave their friends to make sure other people are safe. They really care about people. They have a deep compassion for others, and they'd risk their lives for them. They'd do it in a heartbeat.

"I think the essence of love is laying down your life for others. I really want people to know there are very caring, loving people out here looking out for them. If you've ever felt that nobody loves you, just look at a fire truck going by. Then you'll know that somebody does."

FDNY
YORK STATE
MEDICAL
TECHNICIAN

LIEUTENANT DOREEN ASCATIGNO

AGE 34
DISPATCHER
FIRE DEPARTMENT OF NEW YORK
EMERGENCY MEDICAL SERVICES COMMAND
BROOKLYN

The moment Doreen Ascatigno walked through the door, we were struck by her piercing, intelligent eyes. They were so filled with pain that we immediately put ourselves on emotional guard for the stories that were to follow.

As the interview progressed, we were very moved by the horror and helplessness Doreen experienced in the FDNY Emergency Medical Services dispatch room on September 11. The nightmare of dispatching ambulances to the World Trade Center that day continues to have a profound effect on her daily life.

One of the saddest things we heard while we were in New York City was Doreen's story about the reactions of some of her colleagues to decisions they believed she'd made in the dispatch room. Despite the added pain of those comments, Doreen has found ways to rise above the criticism and actively make a difference in the recovery efforts at Ground Zero. Since our interview, she has been promoted to lieutenant and assigned to Battalion 18 in the Bronx.

"This is pretty much the first time I've spoken with non-FDNY personnel about this," she told us during our interview. "People are always wanting to talk to me about what happened, but I just tell them, 'Now is not the time.'"

"No sooner had we started to get some organization than we got hit with the second plane. All we heard were units screaming on the air. 'A plane just hit Tower Two!'"

—Lieutenant Doreen Ascatigno, November 4, 2001

"I was doing two shifts in Communications so I could get an extra day off. I'd worked from midnight to 8 A.M. dispatching one of the Boroughs, and at 8 A.M. I was placed on the Citywide frequency. That's the frequency where we handle any kind of Mass Casualty Incident—anything that's above and beyond the need for two or three ambulances, like a bomb threat or a big fire or a plane crash.

"We have direct lines to LaGuardia and Kennedy airports, so if anything comes over their towers, we hear it automatically. We also have a direct line to the Port Authority Police Department at the World Trade Center. At 8:48 A.M., an officer called up and said to me, 'We have a plane into the World Trade Center. Send as many ambulances as you can.'

"My first thought was a small Cessna. I asked him, 'Is this a small plane or a big plane?' He said, 'It's a big plane. Just get them here.' So I hung up the phone, put the call into the system, and started to transmit it as a plane crash. As I was doing this, the ambulance supervisors and units were coming up on my frequency saying, 'I'm going, I'm going, I'm going.'

"I started to write things down because it became overwhelming to keep track of everybody. The phones were ringing off the hook. I have a screen that keeps the calls in priority, breaks them down into Boroughs, and narrows them down to areas. Manhattan South is where the World Trade Center is, and I saw the whole Borough of Manhattan light up on my screen.

"We have a predetermined plan for certain big areas like the Empire State Building and the World Trade Center. The first arriving supervisor on the scene immediately yelled, 'This is a hardhat operation!' which meant that he required anyone responding to wear a hardhat. First he asked all units to respond to the preset staging area, but when he got out of his vehicle in front of the World Trade Center, he got so inundated with patients that he started requesting units to respond there.

"I was dispatching all these ambulances to the staging area, but some of them were going directly to the front of the World Trade Center to treat patients. I remember distinctly hearing one of the units that ended up missing saying that they had a patient they were transporting to the hospital. They actually treated somebody and left, and then came back and got caught in the fall.

"I guess as each supervisor arrived, they were getting bombarded. At this point, I couldn't picture it. I had no TV, nothing to see what it was like down there. There's a TV in the supervisor's office, but I couldn't even get up to look. The calls just kept coming in and coming in and coming in.

"We were still thinking that this was a situation where the navigation went bad. No one was thinking that it was a terrorist attack. We were trying to set up staging areas so the ambulances could start transporting patients, but they were getting flagged down on their way in. People were waving them down and they'd have to stop and treat them right then and there.

"No sooner had we started to get some organization than we got hit with the second plane. All we heard were the units screaming on the air. 'A plane just hit! A plane just hit Tower Two!' That's when we knew this was something very big. This wasn't just pilot error.

"The fire department and our chief set up a treatment area in the lobby of Tower Two. Again, we started to get things somewhat organized. We were calling people from home to come in because we thought we were going to get overrun with patients like we did at the World Trade Center bombing in 1993, when we treated thousands of people. We figured we needed to get as many ambulances down there as we could.

"There were so many units responding. It was during a tour change, so people who were getting off work stayed on. We usually ride two to an ambulance, but they were loaded with four or five in the back. People who weren't supposed to be there were there, and we couldn't keep track of them. Even if we had an ambulance count, we definitely didn't know how many EMTs and paramedics were down there.

"We started again, treating and transporting, treating and transporting. We began to get a routine pattern again, when we heard screaming. 'The Tower's coming down! Everybody get out!'

"And then, I think the eeriest thing for me was that there was dead silence.

"I thought at first I'd lost my radio, because we have repeater sites at the World Trade Center. I immediately checked those to see if that was the problem, but they were fine.

"All of us just stood there and waited to hear somebody come back up on the air. At this point, we thought we'd lost everybody.

"We have an alert tone that's supposed to get everybody's attention over the radio, and all I could do was send that out. I didn't know where anybody was. I couldn't get outside help to them, so I figured if they could hear each other's radios, maybe they could help each other.

"All this time, I was listening to the fire radio, I was listening to the police radio, and all I heard was screaming. 'We've got a unit trapped!' There was a chief trapped in Seven World Trade, and he didn't know how to get out. The feeling of helplessness was overwhelming, because I didn't know what to tell them. I didn't know where help was coming from.

"Soon after that, we heard that Tower One was coming down. Again, that eerie silence. I was saying, 'This isn't happening. There's no way. No way, no way, no way.' All I could do was put my head in my hands and ask myself, *What do I do now?* I'm usually in that room alone, but by then, there were maybe four or five other dispatchers in there, most of them answering phones. I turned around to look for advice from my supervisors, and they said, 'There's nothing we can tell you. You've got to do what you feel is best.'

"When the units finally started coming back on the air, I heard seasoned chiefs and supervisors out of breath as they were running. 'I gotta get out of here!' I had a captain saying she was hurt. We didn't hear from her again until the next day.

"Units started coming up on the frequency and asking where they could bring the injured. I kept saying, 'I don't know. I don't know.' It was such a helpless feeling. They finally set up a triage area near the Staten Island Ferry, but the one thing I noticed was that they weren't transporting many patients. We pretty much knew by then that people were mostly dead. We had units calling up saying, 'I'm driving in, and I don't mean to do this, but I'm driving over body parts.'

"At that point, I think our priority was just to see who we had, who we could keep track of, who was missing, and who wasn't. I finally got up and took a break, and that was the first time I actually saw it on TV. I watched the replay on the news, and I felt like I was just going to pass out right there. I couldn't imagine that there were any survivors.

"I felt like I was the one who sent those people in there. I mean, it was under the direction of a supervisor, but I kept questioning myself. *Should I have done this? Should I have done that?* I guess I wondered, *If there had been somebody else in that seat, would things be different? Would they be better? Would they be worse?*

"Before I knew it, they were sending me home to get some rest because they needed me back that evening. Somebody drove me home, and I remember riding in the back seat, just staring at the smoke. To drive on the Brooklyn–Queens Expressway and have nobody else there, to have the whole road to yourself, to see a sign that says, 'City Limits Closed,' is a horrible feeling. Horrible. In New York City. Who would have thought?

"So I went home, showered, packed a bag, and went back in. I didn't get any sleep whatsoever. For 12 days straight I worked on that Citywide frequency, and it was pretty quiet for such an event. Just a matter of removing bodies. I put in my request for a transfer to the field a day after the Towers fell. I requested Manhattan.

"During this time, I felt very guilty. I spoke to people in the field, and they said, 'You did what you were supposed to do.' But every once in a while, I'd meet somebody who didn't know I was the Citywide dispatcher. Just today, as a matter of fact, I was talking to a captain who was saying that no one expected Citywide to send them to West Street. How am I supposed to take that? Is he blaming me? Another person said to me, at a firefighter's funeral, 'Oh, you're the one who told everybody to stay there when the buildings were falling down.'

"It's just an overwhelming feeling of guilt, even though I know deep down that I was just doing what I was supposed to do. I get angry; I get sad. I don't know how to defend myself. Everybody says, 'You did what you could.' The instructor who trained me to be a dispatcher came in to thank me for doing a great job. She said, 'You were calm; you didn't scream; you did whatever you could in the given moment.'

"After the first week, I asked the captain if I could hear the tape recording of that day's dispatches. I sat back and listened to it, and then I listened to it again, trying to see if I did anything wrong. At that point, I knew who had gone in and who was missing, and it gave me chills to hear that unit say, 'I'm responding.' I think I needed to hear it, though. I needed to hear that I didn't send anybody in where they were told not to go. I didn't tell anyone to stay there. I think anybody who was there that day, anybody who had anything to do with it, will always question themselves.

"A week or so later, I went down to the site to start working there as an EMT on my days off. I guess I'm doing that to counteract the helpless feeling I had that day. I want to be there when they find one of us. We lost two FDNY paramedics and another six paramedics and EMTs from the hospitals and private companies. To me, it doesn't matter what patch they wore. They were all there to save lives.

"Down at Ground Zero, we have ambulances for transport, a temporary morgue with personnel to track what comes in, morgue units that transport only members of the fire department, police, and EMS. I've pretty much worked it all. When I worked the vehicle that goes onto the pile, we actually removed nine people, mostly Port Authority personnel, civilians, and one firefighter. I've treated workers who came with injuries and transported them to the hospital. I stand by, and I wait and I wait, for 12, sometimes 16 hours at a time. I don't want to see anybody get hurt, but I'm there, just in case.

"I think the only thing that's keeping people from really losing it is the fact that nothing is identifiable. Most of it is things like a boot with a bone in it or a piece of flesh the size of a stick of gum. I had a hand with a bracelet on it. The one thing I found that was amazing was a jogging suit, still on the hanger, completely intact. I wondered, *How did that not get touched? How did that not get burned?*

"One day, I stood on the edge of the pile looking down when they were removing a firefighter, and I was amazed at how the firefighters and construction workers can go down in there and not get hurt. It's just twisted metal. No computers, no phones, nothing. The only reason they're finding these firefighters and knowing it's them is because of their turnout gear. Their coats don't burn.

"I'm very moved by the way they bring out the firefighters and the police officers, how all of them, no matter what they're doing, even the heavy crane operators, stop and salute. When I'm involved in removing somebody, whether it's a firefighter or a bag of remains, I treat it with respect. Like it was somebody I knew. And it could be, for all I know. It's important to me that family members know I didn't take that bag and fling it. It may sound silly, but it gives me a purpose. Maybe not a lifesaving purpose, but a purpose.

"What's sad is that, even after everything is over and done, there's going to be a lot more death from it. I can see it on the firefighters' faces when I'm down there. One day I had somebody jump into the harbor. He was a battalion chief's son. They pulled him out of the water, and he was just crying, crying, crying. I wanted to hug him; I wanted to hold him. I'm afraid there are going to be so many suicides. I spoke with one of the EMS lieutenants today, and I could see in her face that she's never going to be the same.

"I know I'm not going to be. I know I definitely need to speak to somebody. Even though we're on the other side of the radio, we still need some help. We had Call Receiving Operators telling people on the higher floors above the fires, 'Stay there. They'll come get you. Stay there.' I know one operator who was watching TV and talking with someone on the 90-something floor and saying, 'You'll be fine. You'll be fine.' Then the Tower fell and the line went dead. She cried like a baby.

"I think those operators should definitely not be forgotten. They did a hell of a job trying to talk to these people, to keep them calm. I guess by looking at it, there was no hope for them. The staircases were wiped out, and you could see by the smoke that there was no way for a helicopter to get to them. No way.

"I've become very depressed and shut down. I don't want to talk. I don't want to do anything. I come home and sleep and go back to work. I'm very short-tempered, and it's become an issue with my family. I wake up in the middle of night, and I can't go back to sleep. Most of my nightmares have been about trying to help people and not being able to. I haven't cried yet, but I know it's coming. A part of me died with everybody else that day."

PAPD

SUE KEANE

AGE 42
POLICE OFFICER
PORT AUTHORITY OF NEW YORK AND
NEW JERSEY POLICE DEPARTMENT
MANHATTAN CRIMINAL COURT

Port Authority police officer Sue Keane told us that she comes from a long line of strong women.

"My great-grandmother, my grandmother, and my mother are all very strong and capable women," she said. "I've never known any of them to panic. I guess they must have passed along that backbone to me."

Listening to Sue's account of September 11 certainly didn't convince us otherwise. We were amazed at the number of times she skirted death that morning. With 13 years of military service and 8 years of experience as a police officer, Sue was well prepared to react under the worst of conditions. Her clear thinking and refusal to panic helped save her own life and the lives of many other people that day.

By the time we interviewed Sue, four months after the attack on the World Trade Center, her memory had become a lot clearer about what had happened. Even so, she still wasn't sure exactly where she'd been, what exploded, or which buildings were falling around her as she searched for safe ways to get people out of the complex.

"Every once in a while," she said, "I'll remember something else. I'll remember someone I saw or something I did. I don't wake up from nightmares anymore. I just call them 9/11 dreams. I'll probably have those for the rest of my life."

"It started to get dark, then all of a sudden there was this massive explosion. We were on the mezzanine, which is all encased in glass. The windows blew in, everything went black, and we all got thrown."

—Sue Keane, January 26, 2002

"Anything that connects New York and New Jersey is maintained, run, and controlled by the Port Authority of New York and New Jersey. We're responsible for the George Washington Bridge, all the Staten Island bridges, the Lincoln Tunnel, the Holland Tunnel, the Port Authority Bus Terminal, and all three airports. We have an interstate compact and we work under both governors. We have police powers in both states.

"The history of the Port Authority Police Department goes back to the days of prohibition, when bootleggers would run liquor from New York into New Jersey and turn around and say, 'Well, ha ha, you can't touch us, we're in another state.' The Port Authority got wind of that, and their toll collectors became police officers. Today, we police all the facilities that the Port Authority runs.

"I've been a Port Authority police officer for eight years. I spent most of my time at the bus terminal, but I've been working as the Port Authority Court Liaison Officer for the NYPD for the last two years.

"On September 11, I was at the court building about five blocks away from the World Trade Center. I went outside and got my coffee at the coffee wagon, and I heard a plane that sounded like it was flying really low. I looked up, and all I could see was this big orange fireball at the Trade Center. I dropped my coffee, ran back in the building, called my command, and said, 'Something just happened at the Trade Center. I'm going over there.'

"The World Trade Center is where the Port Authority Trans-Hudson (PATH) trains come in from New Jersey. We have two commands at the World Trade Center, as well as many offices, so I knew a lot of my colleagues would be there. There's a park off Center Street, and I cut through that and got there within minutes.

"Things were starting to drop down from the building—airplane pieces and people. I looked up, and I could see the end of the plane sticking out of the Tower. I was on active duty with the Army for 13 years, and the military trained me not to panic. I was back and forth to Saudi Arabia as a mobilization sergeant for the 77th ARCON Unit during Operation Desert Storm in the early 1990s. I was in Lebanon in 1980 after the Marine barracks were blown up. Things like that prepare you for times like this. You assess the whole situation before you panic, because once you panic, that's it. You're gone. You're useless.

"I decided to go up to the Port Authority Police Desk to see where I could help. People were running past me. They were looking at me like, *Why are you going that way when we're all going this way?* Other people were running up to me saying, 'I'm an EMT. Where do I go?' I had to say, 'I don't know. Don't follow me. Don't follow me.'

"The Police Desk is in Five World Trade. I came in from Church Street, where they were first starting to evacuate people. I saw Sergeant Bobby Kaulfers, one of the only people I ever let call me 'Susie,' and he said, 'Hey Susie. Where are you going?' I said, 'To the Police Desk.' He said, 'Okay. Go on.'

"It's hard for me to talk about this, because we lost Bobby that day. I keep looking back like, if I had been with him, if I'd stayed with him, I wouldn't be here. There were a couple of other people I saw during this time, and I came to find out later that they're gone, too.

"I'm only familiar with Building One, so they told me to go over there and help evacuate people out of a stairwell. At that point, there was already debris in the plaza from the plane and people jumping out the windows. They said the heat was so intense up there that people just couldn't take it anymore.

"It was a beautiful, gorgeous day. I was in a short-sleeve uniform shirt with my uniform pants, my regular shoes, and my side holster. I had a pair of handcuffs and keys. I didn't even have my gun belt on. Later on, I realized that not having it on probably helped save my life. A lot of people who had equipment on got caught on things.

"I crossed the plaza and went over to Building One. There was a stairwell there, Stairwell C, and people were coming out. These people were amazing. They'd all learned how to evacuate since the bombing in 1993. The women had sneakers on, not high shoes. They had masks and flashlights. All the sprinkler systems had gone off, so there were a lot of puddles, and everybody was wet.

"Three or four Port Authority police officers showed up from our other commands. I'm telling you, people from our commands got there so fast. It was like a mass mobilization. Another guy was there who looked like a building engineer, and there was a guy from the FBI and a captain from the NYPD. We were all directing people out, and the firefighters were coming in.

"I kept hearing this whistling noise, so I turned around and saw the bodies coming past the windows. I thought, *Oh, dear mother of God.* The first one I saw had to be female, because she had on a flowered dress. At that point, I think I shut my mind down to it.

"When a body hit the ground, it was a 'thump' with no echo. That's how I knew it was a body. I knew that from training with the military. I could also hear the breaking glass when the bodies hit the skylights on the lower part of the building. Someone told me later that when people jump from that high, they usually have a heart attack and die before they hit the ground. I didn't hear anybody coming from above screaming. People on the ground were screaming, but all I heard from anyone who came from above was a whistle and a 'thump.'

"I said to the other officers, 'Don't let anybody near the windows.' I didn't want anybody to panic. People would try and look out, and I would say, 'No, stay against the wall.' I got very authoritative. 'No. You're going this way. Keep going. Don't worry about anything else.' Usually when you evacuate a building, people will argue with you, but the majority of these people were absolutely fabulous. The people who got out of there deserve a round of applause. They followed everything we told them to do.

"There was a woman who was having an anxiety attack, and we brought her over to the side and tried to get her to calm down. She was saying she was having an asthma attack, and I asked two of the officers to take her downstairs because she was just getting hysterical.

"My survivor guilt is astronomical, because those two officers ended up missing. I know it's not my fault, because we were all just doing our jobs. I know I didn't send them into the burning fires of hell, but I did send them downstairs with that woman. Everybody keeps telling me, 'You can't think that way,' but it's part of my reason for not sleeping half the time.

"All of a sudden I heard the second plane. It sounded like it was coming full force. We were in the stairwell trying to hurry people along, and we heard it hit. Over the radio, I heard, 'Tower Two just got hit.'

"We continued to evacuate. The people coming down the stairs were exhausted. They were having breathing problems. I started seeing people from our medical office, which was on the sixty-third or sixty-fourth floor. I was so happy to see them, because it meant we were getting people out from that far up in the building. Every once in a while the people would stop coming down, and we'd go up the stairs. I made it up as far as the sixth floor, but I couldn't get any farther. I didn't have bunker gear, and the jet fumes and the smell of the fuel were too much. The water running down the stairs was black.

"I think the good thing about that day was that it was Election Day. A lot of people weren't there yet, because they were going to stop and vote on the way in. It was also the first day of school for some kids, so a lot of people were coming in late. Had it been 11 o'clock, this would have been 10 times worse. The observation decks would have been open, a lot of tourists would have been there, and it could have been a lot more devastating.

"There was an older man in a brown suit with a briefcase. I'll never forget him. When I got him out of the stairwell, I stood him by the wall and said, 'I want you to relax for a minute before you leave,' because he looked like he was going to have a heart attack. I was thinking, *I don't have time to do CPR here. I've got to get everybody out.*

"A couple minutes later, it sounded like bombs going off. That's when the explosions happened. I could hear it coming, and I knew something was going to happen. I braced myself. It started to get dark, then all of a sudden there was this massive explosion. We were on the mezzanine, which is all encased in glass. The windows blew in, everything went black, and we all got thrown.

"In the military, they blow things up around you so you're not afraid of it. You don't panic. If something blows up, we cover, we breathe into our shirts, we wait for it to pass over, then we look up before we get up. All that stuff stayed with me. I jumped on the old guy to protect him, and I remember thinking, *If he wasn't already having a heart attack, I just killed him.*

"There was this incredible rush of air, and it literally sucked the breath out of my lungs. Everything went out of me with this massive wind. I instinctively pulled my shirt over my face like a mask. Stuff was just flying past. Then it stopped and got really quiet, and then everything came back at us. I could breathe at that point, but now I was sucking all that stuff in, too. It was almost like a back draft. It sounded like a tornado.

"When the explosion happened, there were officers in the stairwell. The door slammed shut, and there was something blocking the front of it. The guy in the brown suit got up and he was okay, and the captain grabbed me and said, 'C'mon, let's get out of here!' I turned around and heard the officers in the stairwell banging on the door. There was a buckle in the doorframe, and I could see their flashlights through the crack. I pulled myself away from the captain and said, 'No, I've gotta get back over there.' The captain left with the man in the brown suit, but all these people were trapped in the stairwell, and I wasn't going to leave them behind.

"It was pitch black now, and I followed the wall back to where I could see the light behind the buckle in the doorframe. I slid my back down the wall and took my feet and pushed whatever was blocking the door out of the way. I've always felt that women are stronger in the legs than we are in the arms, and I know that's true for me. There's also that 'mother strength,' where in emergency situations mothers have been known to pick up cars off their kids. So I just slid down the wall and used my legs to push away whatever was there.

"There was no handle on the outside of the door, so I yelled through the door, 'Go ahead! Open the door! Open the door!' They pushed the door open, took one look at the blackness behind me, and pulled me into the stairwell. Someone told me later that the explosion sounds I'd heard were actually the cement floors hitting each other when Tower Two was coming down. I never in a million years would have thought the building would fall. I never would have thought that Tower Two had been compressed into seven floors.

"People were still coming down, and we walked up the stairs to see if we could find anyone else. By the time we came back down and opened the door again, everything looked white. It looked like a massive dustbowl out there. It looked like a war zone, like what you see on TV happening in Israel and Ireland. People were still coming out, and we were all shaking off our shock. Then it got really quiet again. I don't know if anybody else felt it, but when it got really quiet, I looked around and said, 'This isn't good. Something's going to happen.'

"Then everything started turning black again, and I heard that noise again that sounded like a railroad train coming. I told everybody, 'Get out of here!' and that's when the second explosion happened and I got thrown and separated from the other guys. I turned around, and all I saw was a beam falling. Or maybe it was a floor. I was on one side of it with a couple NYPD officers and some firefighters, and the Port Authority guys were on the other side. At that point, you couldn't tell what was what. There was paper flying and copy machines falling. And that's when the body parts started to fall.

"I don't know if that was when One World Trade started coming down or if maybe some other building had come down. I really don't know. The timeframe is totally out of whack for me. Again, I got up and started shaking myself off. Every time I got thrown, I felt like I landed on a cloud of air. I never got hurt. When I got to the hospital, any blood that was on me wasn't mine. I was bruised, I had a few scratches, but I was not cut at all. I did a point-check on myself. *Okay, I'm here, I can stand up, I can move. Okay, I'm good.* And I moved on to whatever I had to do next.

"I'm only assuming that Building One was still up at that time. There were still some people there, and there was a doorway to the outside toward Building

Five. We had to look up before we went out because body parts and things were falling from above.

"It's amazing how civilians and people from other departments just clicked and worked together. Standard Operating Procedures and all that crap just went totally out the window that day, and it was like, whatever you could do, up to the point that you could do it, you just did it. This guy—I had no idea who he was—said, 'You gotta look up, you gotta let me know when it's clear. I'm only going to let two through at a time.'

"Everything was dusty. I don't know what time it was, but you would have thought it was the middle of the night. I crossed over to Building Five, and I didn't know what I was stepping on. I didn't know if it was a beam, a copier, or a body part. I was hoping that whatever I stepped on didn't squish. I didn't look down; I only looked up to get those people across safely.

"We did that for a good 10 or 15 minutes, then that guy came across and the door slammed shut, and we didn't see anybody else come out. There were two firefighters and a civilian guy there with me, and we were standing outside Five World Trade. I was thinking in the back of my mind that those other Port Authority officers were gone.

"Then I heard voices in Building Five, and I told the firefighters, 'If you go this way down this wall, there's a doorway that will put you into Building Five.' The Police Desk was in Building Five. I knew that every time I came to the Medical Office, I had to go to the Police Desk, get a pass, and go over to Building One. I didn't know anything else about the World Trade Center complex. I only knew how to get from the Police Desk to Building One, and that probably saved my life.

"We started climbing over debris. Everywhere we went, there was debris. There was nothing that looked like a normal building anymore. Absolutely nothing. Somebody said something about the Pentagon being hit, and that's when I went into total military mode. The training the Port Authority gave me helped me a lot that day, but the instincts that came from my military training really took over. I said, 'We're at war.' I unclicked my holster. This was nuts. Everything was happening too fast.

"We went into Building Five, and another firefighter was hollering, 'We can get out this way!' Building Five had an escalator that went down to Vesey Street. We were standing at the top of this escalator, and I saw some guys from the Emergency Services Unit down at the bottom. They were helping a very heavy woman we had gotten out of the stairwell about an hour before. I thought, *Jesus Christ. I saw her two explosions ago.*

"Outside, it did not look like the beautiful day it was. It was so dark that it seemed like it was 5 o'clock, and I know it wasn't. Something broke my watch at 11:30, so I know that from 9 o'clock until at least 11:30, I was in between and around those buildings, still having no idea what was going on.

"All of a sudden, everything went quiet again. I said to the guy behind me, 'Brace yourself,' and again, another explosion. That sent me and the two firefighters down the stairs, on top of the ESU guys. Again, everybody shook off and got up and tried to regroup. We were trying to get this woman up, and she was really heavy and out of breath; I don't even know if she made it.

"I heard the rumbling again, like a locomotive coming. We all jumped on top of each other and huddled up against Building Five. I may have been in military mode, but there was a part of me that was scared to death. I didn't think I was getting out of there alive.

"Then the same thing happened again, where the wind sucked the breath out of me. I held on to everybody and put my face in my shirt. I braced myself, because I knew that whatever went that way was coming back this way. I think they knew it, too, because none of the others moved.

"Whatever came back, came back in my eyes. We got up and were trying to get the woman on a backboard, and I said, 'Guys, I can't hold on, because I can't see.' I felt useless, because I was tired and I was getting sore. I can't tell you how many times I got banged around. Each one of those explosions picked me up and threw me, but I never got caught on anything, which was amazing. I had to have had a guardian angel in there with me.

"There was another explosion, and I got thrown with two firefighters out onto the street. I looked over where the guys had been helping that lady, and there was no one there. I didn't hear them, and I couldn't see the lights on their helmets. Nothing. To this day, I still don't know what happened to them.

"We were outside, but you couldn't tell the sidewalk from the street. There were smashed fire trucks and police cars. It literally looked like a war-torn country. We started climbing over debris, and I remember climbing over a white Explorer, then climbing over a fire truck. We broke the window out to see if anybody was in it, but nobody was. It was half buried in debris.

"I got stuck in debris twice, and the firefighters grabbed me and pulled me up. If there was a soft spot, I sank into it because I was smaller than they were. As I was climbing out of all this stuff, I saw one of my Port Authority classmates, and we stood there for five minutes hugging each other. There were all these little fires going on around us, but they were minuscule compared to what we'd just been through. The fires at this point meant nothing.

"I started walking down Church Street, and I thought I was in Belfast. All I could hear were sirens. At this point, my body was probably going into shock. My mission was to find Port Authority officers. All I saw were firefighters and NYPD and EMS. It was like *The Twilight Zone.* You come out of this thing, and you're thinking that you're the only one there. Every time I looked at my watch it said 11:30, and it was like time had stopped and I was the only Port Authority officer left.

"What was also going on in the back of my mind were my two kids. Now I was coming out of military mode, and I was going into single-mother mode. I was on a mission to get a hold of my kids to let them know I was okay. I was thinking, *Oh my God, everybody must think I'm dead.* Every time I saw somebody, I asked, 'Do you know where the Port Authority is?' Half of them didn't know what the hell I was talking about.

"I looked down, and my shield was loaded with debris. You couldn't even tell I was a cop. Someone had given me a bottle of water, and I started pouring water over my shield and my patch. I guess I figured that if I dropped dead in the street, they'd know who I was.

"Every time I saw a firefighter, it just killed me, because I kept thinking back to all the ones who went up in the stairwell. I don't think any of them came out alive. I was also thinking about the officers I lost in the building. I was thinking, *I've got to find the Port Authority to tell them that these guys were in the building with me and I lost them.* I found out later that they actually made it out.

"People were being nice to me, wanting to help me, but I wouldn't let anybody stop me because I was on a mission. Someone from an ambulance tried to wash my eyes out and gave me a patch to put over my eye. A firefighter grabbed me and said, 'Are you okay?' I said, 'Yeah, but I gotta find the Port Authority.' They had a hose going, and he threw me under this hose, which in a way felt great, because I didn't realize until then that my skin was actually burning.

"I had burn marks, not like you'd have from a fire, but my face was all red, my chest was red, and for three or four days there was stuff coming out of my body like you wouldn't believe. It was like shrapnel. It's still coming out. I'll look over and have this little pinpoint under my skin and it will be bleeding. My doctor says, 'Welcome that. Your body has its own natural immunity, and it's going to reject anything that doesn't belong there,' which it did for days. I was coughing up black stuff, and there was black stuff coming out of my ears and my nose. There was so much stuff in my eyes.

"I saw a Port Authority lieutenant, and he took me to the Command Post they'd set up. There was an ambulance there, and they stuck me in there with two firefighters. I found out then that it was 1:30 in the afternoon, so for the last two hours I'd just been wandering the area. And I wasn't the only one. We were all bumping into each other—firefighters, police officers, and EMS workers. It looked like *The Night of the Living Dead.* We were all just walking around saying, 'Which way am I going? Have you seen this one? Have you seen that one?'

"They took me to St. Vincent's Hospital, where they were very well equipped. They had neurosurgeons standing by for head wounds and ophthalmologists for eye and face injuries. They had been through this before with the 1993 bombing. As soon as I got out of the ambulance, I heard one of the eye doctors yell, 'Get me her, now!' They took me into a really dark room, and I had three guys staring into my face and looking into my eyes. I could hear them mumbling, 'What's in her eyes?'

"Then another guy came up to me and said, 'I'm a psychiatrist. I'm going to take your gun.' I put my hand on it and said, 'No, you're not.' A cop came over and took my hand and put it on his shield and said, 'I don't know how bad your eyes are, but I really think that if they're going to be doing stuff to you, I should take your gun.' I was feeling with my hand that he had a shield, and I said, 'Okay, you can have it.'

"I finally got a hold of a friend and asked her to call my kids and my mother to let them know I was okay. Another officer from the NYPD came in and sat with me and rubbed my head and asked, 'All you all right?' He just sat there with me. He stayed as long as he could.

"I was home for about 10 days, and I was kicking out the walls. I wanted to go back to work. I wasn't sleeping, because I was worried about everyone else. I had the classic symptoms of post-traumatic stress syndrome and survivor guilt. I was obsessed with vacuuming my apartment every day. I did not want to see a speck of dust. I was doing a lot of pacing. I didn't like being closed in. If you wanted me to go down into a basement, good luck.

"I think people need to know that the Port Authority Police Department was there that day. The World Trade Center was a Port Authority facility. Even though the fire department lost more than 300 members, we actually took the biggest hit percentage-wise. We lost 37 officers. We had probably 1,000 officers there in the first few hours that day, and we lost 37. That's a big hit for a department our size in the span of a day. Plus, we lost a lot of Port Authority employees. We lost 74 people altogether, including the police officers.

"One of the people we lost was Captain Kathy Mazza. I remember many times we had to close the building because of bomb threats, and she was absolutely fantastic. She kept her head about her, and I admired her for that. I never saw panic in her. And always, after every situation, she walked around and thanked everybody. She never forgot to come up and say, 'Thank you. You did a great job. I know it was hard, I know you had to stand here for five hours and you didn't get a break, and I really appreciate it.' Kathy Mazza was a great woman to work with.

"I've always worked in a man's world. I've worked in the military where there were 10 females and 250 guys in a unit. I've always adjusted to it. But sometimes I feel that people don't give women the credit they deserve. I think we react faster and think much more quickly on our feet. People will argue with male cops, but a woman can walk in and defuse a situation before it escalates into something else.

"I know guys who are Vietnam vets who say that September 11 was worse than Vietnam. These are guys you'd never see cry, and I watched them cry. I've seen more men cry since September 11 than I ever have in my entire life—even in the military, with Lebanon and everything else. I've seen a lot of guys cry over this. They've lost their partners and their friends. We all did.

"They always say, 'God takes the good ones,' and in this situation, he did. The officers we lost were terrific. You would trip all over yourself to work with these people. I admired their work ethic, the way they did things, and I think about them a lot.

"I live alone, and I try to occupy myself when I get home. I probably sleep on average two hours at any given time, and I won't sleep upstairs in my bedroom. I haven't slept upstairs since this happened. I sleep downstairs on a futon, just a few feet away from the door. I have to be able to get out. I'm still in escape mode. I have to make sure this will never happen to me again."

OLSZEWSKI
FDNY

CAPTAIN JANICE OLSZEWSKI

AGE 40
EMS CAPTAIN
FIRE DEPARTMENT OF NEW YORK
EMERGENCY MEDICAL SERVICES COMMAND
FDNY HEADQUARTERS
BROOKLYN

When EMS Captain Janice Olszewski told us her story, she spoke in a slow, modulated voice that held us in suspense all the way through to the end. We were amazed by her ability to articulate all the thoughts that ran through her mind in the critical seconds that meant the difference between life and death on September 11.

We were also impressed with the way she made order out of chaos when she arrived at Ground Zero shortly after the first Tower was hit. Even after narrowly escaping death herself, she remained committed to her role as an officer and continued to supervise emergency medical operations at the site.

"The irony of this is that in all my 14 years in the field," she said, "I barely got a scratch on me. Then I take a desk job and almost get killed."

Janice had a hard time accepting the fact that she was alive in the weeks following the World Trade Center tragedy, and she told us that life seems so much more fragile, so much more valuable to her now. She also has a new appreciation for the people of New York City.

"There's a hole now in the skyline," she said. "I feel violated. The whole city feels violated. There's a void in the city, and it's a bad feeling. But it also feels like the whole city has been through this together. It's made us like a tiny little town, like Main Street, USA, where everybody cares about each other."

"I felt myself suffocating. I was more than 100 percent sure I was going to die. Just absolutely convinced and positive I was going to die. There was nowhere to go, there was nowhere to turn, and I was choking to death."

—Captain Janice Olszewski, October 31, 2001

"I'm a captain in the Fire Department of New York's Emergency Medical Services Command, and I answer directly to the Chief and Assistant Chief of EMS. I'm the next one down in the chain of command in EMS Operations. We're responsible for the day-to-day functioning of the EMS service, which includes the deployment of ambulances, staffing, personnel, deciding how many tours to put out for the day, everything. Depending on weather and conditions in the city, we run about 570 tours a day. My department is responsible for making sure that that—and all the minutiae that goes along with it—happens.

"I've been on the job for 14 years. I started as an emergency medical technician, worked as a paramedic for four years, and was a field lieutenant for seven years. I just recently got promoted to captain. I had just come out of a three-week captain's class, took this new administrative position at FDNY Headquarters in Brooklyn, and then four days later, it happened. I didn't even know what to do here yet. I was just getting situated with this position.

"I was at my desk on the morning of September 11, and I'd just gotten my coffee when people started running by me. It was my first week here, and the atmosphere is kind of marble-and-brass corporate, and I thought, *Running? People don't run here.* It was very strange that people were running.

"At that moment, my beeper went off and the text message said, 'World Trade Center hit by plane.' I thought, *Okay, that's why people are running,* but I sort of halfway didn't believe it. I thought, *I'll just go look out the window at the end of the hallway.* There's a perfect view of the World Trade Center from that angle, so I ran down the hall, looked out, and saw the first hole in the building. There were flames and smoke shooting out of it, and it looked huge from where I was standing. I did a classic double take.

"I said to myself, *All right, I gotta go.* I wasn't assigned and I didn't have to go, but I could see that this was huge. It wasn't a small plane that had just nicked the building; something had plowed into it. I was thinking, *I've got to get down there.*

I'm a supervisor. They're going to need all the help they can get. What good can I possibly be sitting behind a desk today?

"I'd just been issued a marked FDNY sedan, and I said to one of the other lieutenants, Bruce Medjuck, "Let's take my car and go down there.' He was already gathering some portable radios.

"Well, I had nothing. I had just started here. My gear, my helmet, my turnout coat—everything was in my personal car back home, and I hadn't had a chance to bring it in yet. I had on a short-sleeve uniform work shirt, and Bruce had some portable radios. That was it.

"All I knew was that I had a terrible feeling of butterflies in my stomach. I thought this would be really, really bad, that we were going to have a lot of injuries on those floors that got hit. I thought a lot of debris would have fallen down and hit people below. I certainly was expecting a lot of chaos, a massive response of all sorts of vehicles, and maybe some panic in the streets. But in the car on the way, I couldn't think of much, because it was an obstacle course trying to get there through the traffic and people. There were all kinds of emergency vehicles starting to head over there, and I didn't have lights and sirens on my car.

"We had to cross the Brooklyn Bridge, which had already been closed to civilian traffic, so we pretty much breezed across that and made our way down there. Bruce was trying to get something on the portable radio, trying to log us into the system so they'd know we were responding. On our way, the second plane hit the second building. We didn't see it or hear about it. We didn't know that had happened until we got there.

"When we arrived, we saw one of our chiefs and walked over to him. I looked up, and there were two holes with smoke and flames, identical at almost the same level in the buildings. I was like, *Did I miss that out the window? I couldn't have missed that.* Now I thought that maybe a plane had careened off one building and hit the other. It wasn't registering. Not yet.

"I was standing on the street looking up, then I clicked over to autopilot. I thought, *I've got to do my job. Somebody's got to maintain some sort of control here.* I couldn't panic. I had to collect myself, so I sort of shook it off. That was the last time I looked up. I never looked up again. Not because I didn't want to. I was just too busy.

"I thought, *This is the worst thing I've ever seen. It's probably the worst thing I will ever see. It's a catastrophe, it's horrendous, but that's it.* Certainly I didn't expect anything else to happen. I never expected more planes to come in, or bombs, or explosions, and certainly not the collapses. I think it was a defense mechanism for me to think that this was all we were going to be dealing with. *This is terrible, but it's over.* That's where my mind was. Nothing else. No more. I didn't think of my personal safety, never once. I just went on autopilot and did my job.

"My chief sent Bruce and me immediately over to One World Trade to set up a triage center on the corner of Fulton Street and Church next to the Millennium Hotel. We were directly across the street from the Twin Towers. There were cops everywhere, and people were streaming out of the buildings. As they were coming out, some kept running, and those who were injured or had smoke inhalation came over to us. They saw that we were collecting there, that ambulances were starting to come in, so they gravitated over to us on their own or the cops sent them over to our corner.

"It was madness. It was chaos. Ambulances started arriving, and as the personnel got out they looked up, and you could see the looks on their faces, like *Oh my God.* It was hard to control people's emotions, and it was hard to get people to get into the mode. You could see by the look in their eyes that they were overwhelmed, but they were ready to do whatever I wanted them to do. As an officer in any Mass Casualty Incident, you've got to give people direction and they've got to take it. I think they were hungry for that, because they didn't know what to do. There was just so much. There was debris all over—paper and chunks of stuff that spewed out from the buildings after the planes hit.

"We were trying to set up a traditional, sectored triage area. We started to get units coming in from all directions, from all sorts of organizations, including the FDNY, hospital units, volunteers, and private companies. It was hard to control a staging area, because they were sort of freelancing and parking askew. It was very difficult to run something traditional because it was such chaos.

"I happened to be the highest-ranking officer in that area. There was no chief, I was a captain, and that was it. So I grabbed somebody's clipboard, and as the lieutenants arrived, I assigned them each a sector. 'You are staging officer. You are transportation officer. You are triage officer.' I wrote their names down and what their assignments were. Then they went off and tried to do those things, which was hard because there was such craziness going on.

"You have to remember, we're talking about a very short time period. We got there about 9:15 A.M., a few minutes after the second plane hit, and the first collapse was around 10 A.M. We had maybe 40 to 45 minutes there on our corner, trying to deal with this. But we started to deal with it. It was bad, it was crazy, but we were starting to get something going. We had triage tags, so we were tagging people. We had an area for the badly injured and an area for those not so bad. We told people who could walk to just keep going.

"There was a wide range of injuries. We had smoke inhalation, burns, soft-tissue injuries, and broken bones. Somebody even had a seizure in the middle of the street. You name it, we saw it. The people were just streaming out, and some were hurt very badly. There was one woman who was just covered with blood and burns. It was really hard to take it all in and focus on any one thing. I wasn't dealing with individual patients. I could only catch them out of the corner of my eye. I was trying to look at the biggest picture I could and keep things organized. I had to keep those lieutenants moving and try to organize those vehicles that were now almost clogging the streets.

"We were able to start transporting patients, but I don't know how many we sent out. If anybody was doing the tracking, that paper's long gone. That's history. There were also a lot of people jumping in vehicles and leaving without even reporting to us. So who knows? They could have pulled up completely out of my eyesight, loaded up, and left. And there were a couple of other triage areas around the corner and up the street that I had no control over, so I didn't know what was going on with that. We were trying to get as many people into an ambulance as we could—five or six people, depending on their level of injury.

"Just before the first collapse, I was thinking, *We need a bigger area, we need a lot more units, we need city buses to come in and transport these people—we need everything.* I was just starting to formulate a bigger plan. We needed to get off that corner because it was getting swamped. We were going to have to maybe take over the hotel lobby and set up a little hospital area. I was just starting to think, *This isn't going to cut it.* We had roped the area off with some tape, but we were getting overwhelmed.

"About that time, a cop came up to me and said, 'You need to send medics into the building. People are hurt in there, and they can't get out.' I was thinking, *It's smoky in there, there's debris falling in there.* EMS doesn't send people into danger like that, because they're not trained, and they don't have the same protection the firefighters have to go in. We don't send people in if it's that dangerous. Under those circumstances, people might have run in on their own. They might have been sent in because it was such an unprecedented situation, but I said, 'No. It's not safe. I'm not sending my people in there.' Ten minutes later, the building fell down. Any people I had sent in would have been killed.

"We had started to get some kind of foothold with the treatment and transportation. I had quite a few lieutenants, ambulances were coming in, and people were getting triaged and treated. It was pretty much going as well as it possibly could for that kind of incident. We were really involved in our patient care, and I have to say, I don't think anybody thought those buildings would collapse.

I certainly wasn't thinking about it. Never for a moment did I think those buildings would fall down. Maybe if they'd kept burning all day, somebody might have said, 'Hey, we'd better get out of here because they might fall.' But that fast, that soon, with all that was going on—it was a preposterous idea.

"My back was turned to the buildings at the moment the first building started to fall, but I heard it. I heard this indescribable roar. I felt it. It was a deep, loud, rumbling, thunderous booming. I heard it happening before I saw anything, and I was already moving. My body was going. I saw all the people start to move away from the noise. I turned around, and this black cloud as high as the sky was wailing. It had debris in it, and it was coming down and at me. I was so close, and it was coming so fast, I just dropped the clipboard, turned around, and ran like I've never run before. I was in an all-out sprint to get away.

"So I didn't see that building collapsing. I was too close. If you saw it on TV or if you were at a distance, you could see it going down. But I didn't see that. I just saw a black cloud obliterating the building. I was running. I wasn't watching. I wasn't even thinking.

"Lieutenant Amy Monroe was standing right next to me as it was happening. I've known Amy for years. We went through the same lieutenant's class together. She just appeared on the scene, I gave her a task, and she was doing it. Somehow we were right next to each other when the collapse started. So we were running together.

"This I'll never forget. She turned to me and said, 'Janice, I'm scared. I'm very, very scared.' The look on her face was one of such fear that I wanted to pick her up and carry her. I thought, *She's not going to make it. She's too small.* The look on her face was complete and utter panic, and I thought, *She's not going to get out of here.* Later on, I found out she's a black belt in karate, and she probably should have been carrying me. But I didn't know that at the time.

"This all happened so fast. I wanted to run with Amy, but we got separated instantly. The cloud came on us too fast. I knew everybody who had been there with us—20 to 25 people, some of them good friends I've known for years. Every single one of us got incredibly scattered. We just ran.

"Everybody was running. Some people tripped and fell, and others helped them up. It was strangely not a trampling thing, which was shocking, because it easily could have been. Everybody was trying to keep their space so they wouldn't bump into anybody and knock them down. I miraculously didn't fall. My friends said later, 'We thought you were dead because you're such a klutz. We thought you would have fallen and gotten run over by something or buried.' I said, 'No, believe it or not, I did not fall once, for a change.'

"I stayed on my feet, and I had an option to go down into the subway, but in a split second I thought, *No, that's probably not a good idea to go down into a hole with this debris falling and all this smoke.* I thought the lights would go out, people would get crushed down there, the smoke might get sucked in, or debris might cover the hole. I was not going down there. That's a deathtrap. I thought, *If something's going to happen, I'm going to take my chances on the outside.*

"I kept running, thinking that at any moment tons of debris were going to come at us. I thought it was an explosion of some sort. Either the plane exploded from the jet fuel, or there was a secondary bomb. I thought we were going to get hit with something or overtaken by a fireball. I was thinking, *I have no chance here. I'm running, but I'm probably going to die.*

"I was waiting for something to hit me, but nothing big did. The gray smoke came in first, and there was just no outrunning that cloud. When it caught up to me, it got so incredibly dark that I had to stop running. It was so dark I didn't know which end was up. It was darker than night, because I couldn't see silhouettes and I couldn't see lights. It covered everything. I didn't know which way to turn. I had no idea which way to go. I could have been running right back into it, so I had to stop.

"It also got incredibly quiet. Silent. Like when you go outside after a new snow and it's a blanketed quiet. Nothing. It was really eerie.

"All I heard were the alarms that meant that firefighters were down and unable to move. I heard a couple of people asking for help, but I couldn't see them at all. The stuff in the cloud was so thick. It was pulverized marble, concrete, glass, dust, ash, chemicals, and smoke. I couldn't breathe.

"I started to choke, and I wasn't just choking up smoke like dry heaves, I was choking up chunks. I had to get it out, and I couldn't keep up with it. It was in my nose and in my eyes, and it was so incredibly thick, this gunk. I felt myself suffocating. There was nowhere to go, there was nowhere to turn, and I was choking to death. I was thinking, *I'm going to die.* I was more than 100 percent sure I was going to die. Just absolutely convinced and positive I was going to die. For a split second I also thought, *Just let it happen then.* I was resigned to it, which was a horrible feeling. There was a feeling of giving up, because I thought it was hopeless.

"Then I snapped out of it. I had a couple of distinct thoughts. One was my family and another was, *I'm not dying on these terms. Not like this. Whoever did that back there, they're not getting me like this.*

"I was thinking, *I have to stay on my feet, and I have to stay conscious.* I wasn't getting enough oxygen to my brain and I was starting to go unconscious. I thought, *If I go unconscious, I'm going to fall down and suffocate in this smoke. I gotta get out of this smoke somehow.* I stayed conscious, and I stayed on my feet. I was staggering, and I bumped into a car. The car was perpendicular to me, so I figured I was at the next intersection.

"This incredible survival instinct clicked in at that point, and I found the will to get out of there. I was thinking of my little nephews, Tommy and Johnny, who are still babies, and I was thinking, *I want to see them grow up.* At that point, I saw a red traffic light in the darkness. I thought, *Wow. That's a great reference point. I can use that. I don't know what direction it is, but I'm going to follow that light.* The whole world became focused again because I had something to go by.

"So I followed the red light, then I saw another one and followed that, then another. Slowly I made my way out of the smoke. I just followed the lights and felt my way out. It got brighter and brighter and then I busted out into the bright, brilliant, blue sunshine on Broadway. And there I was.

"I was stunned and shocked and my immediate thought was, *My friends are all dead. Every single person I left behind is dead. I am lucky, this is a fluke, this is a miracle. I got out and my friends are dead.* I was convinced, and so was everybody else who was there, that the others were all dead because we didn't come out together. Nobody came out with another person who was there. We had no idea where anyone else was or what had become of them.

"A lot of people thought I was dead. They had written me off as dead, because when Amy came out, she was telling anybody she saw, 'Janice Olszewski is dead.' Like it was a fact. She didn't know I'd gotten out, and she couldn't believe that I had. She was panicked, too, and that's what she believed. Just the way I couldn't believe that anybody else was alive.

"I thought, *I've got to get away from this scene until I know what's going on,* so I continued north on Broadway. Now ambulances were starting to come toward me, the fresh troops coming in after the first collapse. I was trying to stop them from going in, because I still didn't know what had happened. All I knew was that it was something catastrophic and it could still be happening, or it could happen again. I didn't want them to go in yet until we knew. Some stopped, and some didn't. I told those who did stop that we needed to go farther north, that we had to get away from the smoke.

"I was still sort of shocked and staggering, but I was trying to get back into supervisor mode and give some direction. I was thinking, *Everybody's dead. I survived. I'm a supervisor. Maybe I should try to supervise.* And then I heard the rumble again. I guess I was about three blocks away. *Now what?* I turned around, and that frigging cloud came at me again. This time I was able to outrun it easily because I was farther away. I thought, *What's going on here?* I still didn't know the Towers had collapsed. Nobody on the streets knew, either.

"So I was running again, and I got caught in it a little bit, but I kept going north. Again, more ambulances were coming toward me, and I stopped them, too. Some didn't stop because they were probably saying, 'Here's this really tall, crazy lady covered in dust. Why should we stop for her?'

"I got into one ambulance that was full of people who worked at headquarters, and I said, 'Listen. Turn this thing around and just drive north.' They were like, 'But, but, but …' and I was like, 'No. Drive away. Just go as far away from this as you possibly can right now.' So we drove about 10 blocks up Broadway and started to set up another triage area. I gathered about five or six ambulances on Broadway, and we set up an area for people who were coming out of the incident with injuries or smoke inhalation. I didn't want to go back to the site yet, because I still wasn't sure what was going on. I still hadn't seen anybody alive who'd been there with me.

"I was telling people, 'I don't know who's alive back there, I don't know who's dead, I don't know who's in charge, I don't know where the new Command Post is. I don't know anything.' At that point, somebody told me that the Pentagon and the White House had been hit. I said, 'Oh my God, they're getting the whole country.'

"I realized that I needed to call home because this could be it, and I just wanted to talk to somebody. Nobody's cell phone was working. People said later they'd tried to beep me, but my pager wasn't working either. I ducked into a store on Broadway and asked the guy if I could use the phone, dialed my parents' house, and my dad answered. They were watching it on TV, knowing I'd be there, and they were worried. That was a very hard phone call, because I was trying to be a supervisor and maintain some composure with the paramedics and EMTs outside, but once I got my dad on the phone, I lost it. I cried, screamed, ranted, and raved like a madwoman. I said, 'Dad, I think there are going to be thousands of people dead.'

"I went back out and tried to organize patient care with units coming in, grabbed somebody's portable radio, got on the air, and said, 'I'm here, I've got an area here for patients.' Somehow we found out where the Command Post was, and I said to one of the units, 'I've got to go back there because I'm a supervisor, and maybe I can help.'

"I was covered with dust. I was completely coated—my hair, my uniform. I couldn't breathe, I was coughing, I could hardly talk. All the paramedics and EMTs were saying, 'You need to go to the hospital.' I said, 'No. I want to go help.' It was a hard decision to go back there after what I'd just been through, but I figured I had to go back. I still didn't know those buildings had fallen down. Nobody had told me, because they probably figured I already knew. Even my parents didn't tell me that the buildings had fallen, and they were watching it on TV. I was right there. I was right under it, so of course they thought I'd know. But I still didn't know what had happened, and I didn't even think to ask.

"I asked a unit to take me back to the Command Post, but we got caught up in an emergency vehicle traffic jam and were stopped. I got out and bumped into one of my supervisors, and he took one look at me, grabbed me by the shirt, and said, 'Get back in the ambulance and go to the hospital.' He pushed me back in, and I really couldn't argue with him at that point. I wasn't in any condition to function.

"So that ambulance took me to NYU Medical Center, and I thought, *Gee, I'm monopolizing this ambulance when it could be used for five or six patients. I shouldn't go to the hospital, because it's probably packed with people who are injured worse than I am.* We got there, and there was an army of doctors and nurses and staff outside the hospital, inside the emergency room, and all over the place. I was treated by a team of people. Immediately. I looked around and said, 'Where are all your patients?' This was about two hours after the collapses. I asked, 'How many did you get from the incident?' They said, 'Twelve.' I said, 'Twelve? People? From that? Is that all you got?' Then it hit me like a train. I said, 'You're not going to get any more. They're all dead.'

"They treated me for smoke inhalation, gave me some IV fluids, and about an hour and a half later I said, 'Listen, can I go?' and the doctor said, 'Okay.' I put my same dirty clothes back on and left the hospital.

"I was in the middle of Manhattan, wandering. My car got crushed back at the scene, and I had no way of getting anywhere. Nobody was available to come and get me. I didn't know where to go or what to do, so I walked to the nearest ambulance battalion, which was at Bellevue Hospital, and I said, 'I really hate to bother you, and I know you're very busy, but can I get a ride somehow back to headquarters in Brooklyn?' I figured I couldn't go back to the scene, because I couldn't be breathing anymore of that stuff in. It was just too strong, and I have allergies, and I would be no good there. I thought I'd go back to headquarters and work.

"So I sat in the lounge waiting for a ride. There was a TV on, and that's when I found out that the buildings had collapsed. I was in total shock. People were sitting there watching, and I said, 'They fell down? They both fell all the way down?' And they were like, 'Yeah. Where have you been?' I couldn't believe it. I thought, *How could anybody have possibly survived it?* The only reason we did was because the buildings fell straight down. That's the only chance that anybody on the fringe had. Forget it if you were in the building or on the plaza.

"I finally got a ride back to headquarters and staggered up to the office covered in dust. There were a few people there, and they were so glad to see me alive. They found me towels, soap, shampoo, a new set of clothes, and sent me to the shower. All the other workers were still out there, and I didn't know if they were alive or not. I was alone there in Operations, and I took it upon myself to man the office. One of the secretaries stayed with me overnight, but I was the only officer there. I worked until six the next morning.

"Miraculously, most of the people I was with made it out, including Amy. I was grateful and amazed that they made it out. It was hard to believe. We lost a total of eight EMS workers, and one of the two paramedics we lost from the FDNY happened to be right there with me. That bothered me. Why didn't he make it out when the rest of us did? What happened to him? Where did he go? Which way did he run? We had no idea.

"I got old that day. I'm 40, and before this, I felt about 30. Now I feel 45. I feel haunted by it, but I also feel very inspired. I'm inspired by my co-workers who were there that day. I'm inspired by the people of this city, their reaction to it, their response, and their support. I'm inspired by the country's reaction to it, too.

"It's still sinking in. It's still hard to believe it happened, my little tiny piece of this huge thing. For the first couple of days, it was very hard for me to see the bigger picture because of my own little horror. I had to get over my own physical and mental thing first. Then my thoughts started to branch out a little. My brother-in-law is a cop, and I didn't even think to ask what had happened to him that day until later that same evening. I was just so sucked into my own thing and the amazement of my own survival.

"I'm still absorbing it and putting it in perspective. *Okay, that was horrible. I almost died. What a terrible thing to happen to me. But I'm alive, and other people died. So, therefore, let's put that in its place.* That's helped me cope with it. I could be dead. So many other people didn't make it.

"I talk to people about it, and they say, 'It wasn't your time. You weren't meant to die that day.' I believe in fate, so maybe there was a reason I survived. I'm not very religious, but I do feel that something was at work there, some sort of presence or force. I don't want to get too melodramatic or cosmic about it, but I felt like something was going on there to get me out, to give me back the will. That red traffic light I saw was almost like a gift from God. I saw it at just the right time, because I was going unconscious and I was about to give up.

"The women who there that day were so brave, so heroic, so completely and utterly ready to go right in, without question, without thinking of themselves. They were going in there to help without being told to and without being ordered to. They were just going right in. It was amazing, the selflessness of people. Everybody. And certainly the women. They were as heroic as anyone else and willing to do their part. They had families, kids, husbands, parents, and partners worried about them, and they went right in. They wanted to do anything they could do to help. A lot of people did things I can't believe they did, like Amy, who was treated at the hospital and then went right back out on the pile. Where does that come from?

"Putting yourself in the mouth of danger and not knowing what's going to happen next. That willingness to completely disregard your own life to go in and help others. I'm awfully proud of everybody. I learned a lot about myself, my co-workers, and the people of this city. I'm very proud of them all."

POLICE
DEPARTMENT
CITY OF NEW YORK

LIEUTENANT SPECIAL ASSIGNMENT TERRI TOBIN

AGE 39
LIEUTENANT SPECIAL ASSIGNMENT
NEW YORK CITY POLICE DEPARTMENT
DEPUTY COMMISSIONER'S OFFICE OF PUBLIC INFORMATION

When we met with Lieutenant Special Assignment Terri Tobin on the day after Thanksgiving, her head was bandaged and she was still limping from injuries she received in the collapse of the World Trade Center. Nonetheless, she was back at work in the Deputy Commissioner's Office of Public Information, fielding calls and working with the press.

What was most striking to us about Terri's story was the fact that, despite the life-threatening nature of her injuries, she helped dig three people out of the rubble and evacuated a hundred people from an unstable apartment building before she sought medical attention for herself.

"I was down there to deal with the media," she said, "but I was still a police officer. My main objective has always been to help people."

Terri has been a police officer for nearly 20 years and comes from a family of cops. Her father and three brothers are also police officers. Her service with the NYPD includes a number of years as a patrol officer and working to keep kids off drugs.

In November, Terri was one of six women honored with a Liberty Award by the National Organization for Women.

"The force of the explosion literally blew me out of my shoes. It lifted me up and propelled me out, all the way to the other side of the street."

—Lieutenant Special Assignment Terri Tobin, November 26, 2001

"The Deputy Commissioner usually comes in at about 8:30 in the morning, and we begin going over the schedule of events for the day. My job is also to brief him on what's happened overnight. I come in early and go through all the reports; like last night, there was a shooting in Midtown Manhattan, and the press is calling on it.

"Just after we got started working on the morning of September 11, a sergeant came running in and said, 'A plane just hit the World Trade Center! I'm heading over.' Two seconds later the phone rang, and it was Chief Fahey. He said, 'I want a lieutenant there. Send Terri.' So I grabbed a sergeant and we headed over.

"Rather than try to fight traffic across town, I jumped on FDR Drive and swung around to the World Trade Center. As we got closer, it was like a ticker-tape parade. There were thousands of pages flying through the air. I said, 'Wow. It must have been a direct hit.' There was just too much paper.

"Obviously, I knew it was a big event, but the whole time I was going over there, I was thinking that it was this little Cessna from New Jersey. I was thinking that some guy had a heart attack and clipped the World Trade Center by accident. I was expecting to see half a plane sticking out of the Tower, never thinking that it was a commercial airliner.

"We got there at 8:54 A.M. I remember the time, because my text pager had gone off, and it said that the highest level of mobilization had been called for the World Trade Center at Church and Vesey streets. I parked my car on the east side of West Street, and the mobilization was diagonal to where I parked. The World Trade Center covers 16 acres, so it was a mile walk.

"The North Tower had been hit first. The streets were already closed down because of glass and debris. We had to kind of scoot around the outskirts, and as we were doing that, the second plane hit just after 9 o'clock.

"The television images don't give you the impression of how bad it was when the second plane hit. It wasn't like the plane hit and the glass fell. As the building was heating up, all the windows were popping out. You can imagine these huge windows breaking into shards and falling to the ground. I'd be walking along, and *crash!*—glass and debris falling everywhere.

"Then we started to see people jumping. It was horrendous. At first I did a double take, because the buildings were 110 stories tall. I thought, *Oh man, something's coming down.* At first, it was like a little speck. Then I realized it was a human being. People jumped alone and in groups, and, at one point, they said 20 people held hands and jumped. I can't imagine what hell it must have been inside there, with jumping being their only alternative. I can't imagine the fierceness of that fire. I heard later that one guy e-mailed his mother and said, 'Mom, I have to jump. I hope you understand.' In some cases, the windows were gone and people were holding on to the edge. At a certain point, I think it just got too hot and they had to let go.

"As we approached the mobilization area, I saw First Deputy Commissioner Joe Dunne. He said, 'Terri, there's a report of a third plane in the area. Get a helmet on.' There was an Emergency Services Unit truck parked there, so I jumped on the truck and grabbed two helmets—one for myself, and one for the sergeant who was with me. This helmet is made of Kevlar, and it's real heavy. It's made specifically for combat situations, because it can take a bullet.

"After I got the helmet on, I saw a Channel 7 news crew. My job was to corral all the press together and get information out to them as quickly as possible so we could start doing emergency broadcasts. We needed to let the public know what areas were in the frozen zone, which trains were stopped, where people were being directed to evacuate, and all the other mechanical pieces of a disaster.

"I went into the North Tower, looking for press. It was amazing how calm people were. Everything was going very smoothly, because I think those people were familiar with the evacuation drill. Obviously, they knew that the building had rocked, and there was fear. At one point, I know, there was an hour-long back-up to the thirty-first floor, just because of the volume of people making it down the accessible stairways. But everything was going pretty smoothly.

"I'm telling you, everywhere you looked, there was a cop. For people coming out of the Towers, there was absolutely no thinking. They didn't come down and have to wonder, *Where do I go?* It wasn't mass confusion. Every five feet there was a cop saying, very calmly, 'Please exit to your left. Exit to your left.' So there was this very obvious route. Just follow the blue. At the same time, firefighters were coming in and going up. The response by emergency personnel was just phenomenal.

"Everything was under control in the North Tower, so I crossed over to the South Tower, and I saw a photographer. The last phase of descent onto the concourse level was an escalator, and there was a photographer at the bottom of it. He was clicking away and slowing people down. Also, he was in the frozen zone. We were working really hard to get people out, so the last thing we wanted were unauthorized people getting in the way.

"But I also realized that he had a job to do, so I said, 'You've got two shots left.' Then I literally grabbed him by the collar and walked him out to Liberty Street. I handed him off to a uniformed cop and said, 'You need to walk him as far away from the frozen zone as you can. I don't want to see him again.'

"I was thinking, *This is going to be a long day, and there's going to be a lot of running around, so I should put my sneakers on.* I was walking back to my car and I got close enough to pop the trunk with the remote, when I heard a loud rumbling sound. It sounded like a train. I turned around, left my trunk open, and started walking back toward the South Tower. I was thinking, *Where is this train coming from?* It just wasn't clicking with me that there was no elevated train. They're all underground. I guess it was almost like I couldn't see the forest for the trees. I mean, I was right there. If I'd looked up, I would have known that the building was starting to pancake down.

"Then I saw people running toward me, and they were screaming, 'Go! Go! It's coming down!' Just for a second, I looked up and saw it. I thought, *I'm not going to outrun this.* But then I thought, *Maybe I can make it back to my car and jump in the back seat.* Before I could make a move, the force of the explosion literally blew me out of my shoes. It lifted me up and propelled me out, over a concrete barrier, all the way to the other side of the street. I landed face-first on a grassy area outside the Financial Center, and after I landed there, I just got pelted with debris coming out of this big black cloud.

"And then I felt it, but what sticks with me is hearing it: the *whomp* of my helmet when I got hit in the head. The helmet literally went *crack,* split in half, and fell off my head. I realized then that I'd just taken a real big whack in the head. I felt blood going down the back of my neck, and when I was able to reach around, I felt this chunk of cement sticking out three or four inches from the back of my head. It was completely embedded in my skull.

"Then it got pitch black, and I thought, *I must have been knocked unconscious, because it's totally black.* But then I thought, *I wouldn't be thinking about how black it is if I'm unconscious.* And it was really hard to breathe. All I heard were people screaming. Screaming bloody murder. All sorts of cries. At that moment, I thought, *This is it. We're all going to die on the street.*

"All this stuff just kept coming down and was piling up on me. All of a sudden, I heard these huge explosions. *Boom! Boom!* And I thought, *Now they've started bombing.* I was lying there, I was still getting pelted with stuff, I had a headful of cement, and now if they were bombing us, there was going to be no way out.

"I don't know how much time passed, but I knew my eyes were burning, I was coughing, and my nose was running. From underneath the rubble, a short distance from me, I saw the silhouette of a firefighter's helmet. He turned on his flashlight and shouted over, 'Are you okay?' I said, 'Yeah, I'm okay.' He said, 'Cover your mouth and nose with the front of your blouse.'

"I reached around and grabbed someone's hand. I couldn't see him, but at least I had a hold of his fingers. I said, 'I'm with the NYPD, and there's a firefighter on my right. Just stay down!' And then it started to clear. It went from total blackness to what seemed like white ash, but it was actually pulverized cement sweeping through the air.

"I realized that I was buried under all this rubble. I was finally able to move the top part of my body out, but my legs were still buried. I said to this person whose hand I was holding, 'I'm going to try and get up, but I won't let you go.' I lifted the upper part of my body out and I pulled on him, but he came up too easily. I looked down and realized that I just had hold of a hand and an arm. No one was attached to it.

"I went back down and started digging through the rubble, because I thought someone must be there. But I couldn't find a body. I had this arm, but I couldn't find the person it belonged to.

"And I thought, *Is this a hand from one of the plane victims? The arm of someone who jumped, or someone who was running and got hit with debris?* I don't know. And I'll never know. All I could do was say a prayer. I said, 'If this person's alive, I hope the rest of his limbs are intact.' But in all likelihood, that person was gone. So I just said a prayer for his spirit. I hope it was quick. That's about the only thing you can pray for, that it was quick, because it's so atrocious.

"I was sort of kneeling in the rubble, still trying to get the rest of my legs out. Attached to the piece of cement that hit me were building cables—not like thin phone lines, but thick cables—and I was literally draped in them. I had to physically take them up and over me, and it took a while to get untangled and get up. The firefighter came over and asked me if I was all right. Then he yelled out, 'EMS! She's got cement in her head!'

"Two guys from EMS came over, and they wouldn't even attempt to pull the cement out. They just wrapped my head. My hair was singed, I was covered in this white stuff, and I could smell burning hair and burning flesh. The odors were really bad. All around me, I heard people throwing up. I could hear that choking sound of people just wanting to spit out all the blackness, to clear it out of their lungs.

"I was trying to cough all that stuff out, too. I'd just inhaled two minutes' worth of this dark black smoke and all this white stuff, and I didn't even know what it was. I spit out a chunk of what I thought was cement, but that wasn't what it was. The force of being hit on the head had blown my wisdom tooth out, root and all.

"So now there were four of us standing there, and it became eerily quiet. We didn't hear anything. I turned around to look for my car, and it was totally engulfed in flames. There was a fire truck on fire, and there was an ambulance on fire. It literally looked like a war zone. It didn't click until later on, but those were the *booms* I heard earlier. The force of the structure coming down and the heat from the fire caused the gas tanks to explode.

"Then we began to make out really subtle sounds. We knew that there were only rescue workers in that area, because no civilians had come that way. It had been all rescue people who had been running toward me. The blue and black uniforms. I'd be going along and I'd find radios just lying there, and I didn't know what had happened to the people. There'd just be a radio. Fire, police, and EMS. I'd pick it up and try to raise anyone on it. 'DCPI Lieutenant to Central K.' But I'd get no response.

"The first guy we got out was a firefighter who had rolled under an ambulance that didn't explode. As the rubble came down, it just kept piling against it, and he was trapped underneath.

"So now he joined us, and we were listening for voices. It was just overwhelming. Where were the people? I was thinking how lucky I was because I was conscious. If I'd been knocked unconscious, I wouldn't have been able to call for help and no one would have seen me because I was underneath the rubble. That's why, in my heart, I thought they were going to find people who were just knocked unconscious but who maybe had a breathing area through the rubble where they could get oxygen. Even much later in the day, and even the next day, I still thought they were going to find people.

"We got maybe three people out, and all of a sudden, another whole group of people came running toward us saying, 'The other Tower is coming down! Run!' I was in this mental zone, and I was just thinking, *Where am I going to go?* I'm very familiar with the area, so I thought, *I'll run to the water.* I felt very unprotected at this time. The ash was really thick and I was barefoot, and certain things, like the steel beams that had fallen from the first Tower, were really hot. It amazed me that the soles of my feet weren't absolutely ripped apart.

"So I was saying to myself, *I'm going to run toward the water.* There's just a railing, and I thought I'd jump over that and go into the Hudson River. I know that where there's water, there's oxygen. And as the stuff was coming down, I was thinking, *At least the water will get the impact, not my body.* As I was running, I realized that my foot hurt, and then all of a sudden, I got whacked in the back. I went straight down to my knees. I felt that same rush again. That same cloud of swirling debris was coming at me.

"I knew at this point that I was not going to make it to the water, and I saw a building off to my right that I thought I could make it into. So I got up and took my gun out. I figured that if the door was locked, at least I could blow out the lock.

"But the door opened and I went in. I was thinking at first that the building was under construction, because all I saw were white lights in yellow cages. It never dawned on me that they had lost their electricity. So I followed these lights and I got to the elevator bank, and all the elevators were stopped on the first floor. I figured, *Okay, everyone's gone.*

"I'll tell you the truth, I just wanted to get away from the blast. There were these huge windows in the front of the building, and I was thinking, *Something's going to go right through those.* So I opened the door to the stairwell and looked up, and it was lined with residents of this apartment building. People had just come out of the shower, and there were little babies—probably a hundred people.

"I told them we had to get out, because I thought everything was going to come down. I'm sure I didn't look credible. I was wearing torn pantyhose, my head was bandaged, and I was covered in dust. But everyone got out of the stairwell, and I told them to go into the lobby but to stay away from the windows. At least there would be a way out of the building if it started to come down.

"I went to the front door, and outside it had turned to that white ash. It was exactly the same scenario as before. I saw two guys from our Technical Assistance Response Unit, and on the backs of their shirts was 'TARU' in big white letters. I opened up the door and shouted, 'TARU!' They turned immediately, and one of the guys knew me. I said, 'Listen, we've got to get these people out.'

"He told me that they were evacuating people by boat to Jersey, so I said, 'All right, let's get these people down there.' He kind of turned white and said, 'Well, we'll get *you* down there. We'll carry you.' And I said, 'I'm all right. I can walk.' Then he said, 'Terri, there's this windowpane stuck through your back, between your shoulder blades. It's sticking out of your blouse.'

"Obviously, I'd felt it and I knew that something was there, but I didn't know it was glass. All I could think of was what I'd learned in training, that if you go to a stabbing, you just wrap the knife to stabilize it, because the last thing you want to do is pull it out and cause more damage. So I said, 'Just leave it. I'm okay.'

"We started walking with all these people toward the river, and along this promenade was a van. One of the guys said, 'I really think you should get in this van and get to the doctor.' There were a couple of ferries there, and in the distance, I saw the NYPD Harbor Unit coming in. I said, 'No, I'll just wait for the police boat.'

"By then the sun was starting to come out. It was so bizarre. It was an absolutely beautiful day. Now the sun was starting to come out and the smoke was clearing, and even all the white ash and debris were settling down. We were just packing people on these boats, and the officers were saying, 'Get on! Get on! Get on the boat with them!' I was just not thinking straight. All I was thinking was, *I don't want to go to Jersey. If I go to Jersey, how am I ever going to get back?*

"When the Harbor Patrol came in, the captain looked at me and put his hand on my shoulder and yelled, 'EMS! Get over here! We've got an injured officer!' So two guys came over, and the first thing they said was, 'We're going to have to cut your blouse.'

"All I could think was, *I'm going to be in my bra in downtown Manhattan?* They said, 'We have to,' so I said okay. Then one said to turn around and hold on to the wooden railing. He put his foot up on the railing and said, 'Ready? One. Two. Three!' and ripped the glass out of my back. I was thinking, *If he needs that much force to rip this thing out, how deep was it into me?*

"I was saying, 'I think I'd better stay in Manhattan.' And the captain said, 'Look. You don't have time. Ellis Island is set up to get people to hospitals. Really. You've got to go. Just get on the boat.'

"So I got on the boat, and there was another cop who was extremely young. He had this baby face and he was just lying there. As we approached Ellis Island, they ripped off his shirt and got out the paddles, getting ready to defibrillate him. I was thinking, *God. This guy must be all of 22 years old.*

"I really didn't feel like my medical condition was anything pressing. I think that the hand I pulled out of the rubble put that in perspective for me. I was able to walk and talk. I was fine. Meanwhile—and this is one of those stupid things—I was worried about an 11 o'clock meeting I was going to miss. The enormity of the situation just didn't click.

"We pulled into the dock, and as I was getting off the boat and climbing the stairs, I realized again that my foot was hurting. I got into a wheelchair, and I saw a guy with broken legs sitting under a tree. Both of his legs were broken right below his knees. They just made a right-hand turn, and it was so unnatural looking. When they took me up to the ambulance, I said, 'There's a guy over there, and he's got really messed-up legs. Go get him.' So they went back for him.

"In the ambulance, he looked at me and said, 'Hey, I know you.' I said, 'I'm Terri Tobin. I work in the press office with the NYPD.' He said, 'Oh, I'm David Hanscher. I'm a *Daily News* photographer.'

"The road we were on was gravel, and any time there was movement, he had so much pain. I took his hand and held it and said, 'If it hurts, just squeeze my hand. It's going to be a short ride to the hospital, and then they'll put you out and you won't feel this pain.'

"Later, I was able to call his wife. I said, 'Listen, I rode in the ambulance with David, and he's fine. His legs are broken, but he's conscious and alert. He loves you very much.'

"They took us to Bayonne Medical Center in New Jersey, and they immediately sent me up to surgery. The surgeon said, 'The good news is, we're going to cut this cement out of your head, stitch up your back, x-ray your ankle, and you'll be okay. The bad news is, because you've had blunt trauma to your head, we can't give you any anesthesia.'

"So she cleaned out my back, stitched me between the shoulder blades, and took an x-ray of my ankle and found out that it was fractured. I had more than 80 stitches, a bruised kidney, and hearing loss that's probably going to be permanent.

"It's been a little more than two months since this happened, and I don't think it's really hit me yet. I haven't had flashbacks, I haven't had one nightmare, and I have had no problem eating. I've kept myself busy, going for follow-up medical visits, dealing with the death of my cousin who was a firefighter, and going to funeral after funeral. There hasn't been time to think. I came back to work because people here have been working really hard and they needed a break. Although I'm not able to go out digging at Ground Zero, at least I can answer phones so that someone else can go down there, or get a day off, or go to a funeral.

"I was just back to work a week when a plane crashed a block and a half away from my house in Rockaway. My neighbor across the street had one of the jet engines fall in her yard, and there were engine parts in my backyard. In some ways, that was more personal.

"It was body on top of body with that plane crash. We bagged 40 bodies immediately from that devastation. It was tough, because the United Nations was in session, and we had press from all over the world in town covering the UN. They all came down to Rockaway, so that was a bit much. I was doing 18 hours a day out there.

"So life is just sweeping by. I don't think it's all caught up to me yet. I haven't had time to absorb it all.

"Would I have chosen not to respond that day? No. I still would have gone. I think I'm meant to be doing what I'm doing right now. I can't explain to you how I'm alive, being where I was. The fact that I didn't die is beyond me. If I hadn't had that helmet on, I would have been decapitated. The glass went through my back, but it didn't puncture my lung or hit an artery. There are just so many blessings from that day.

"I could retire this spring, but that's not my plan right now. I think that with all those blessings comes a responsibility. There is some reason why I made it out alive. I can't explain it now, but I think it will be revealed as time goes on."

321

CAPTAIN MARIANNE MONAHAN

AGE 44
CAPTAIN
FIRE DEPARTMENT OF NEW YORK
ENGINE 321
MARINE PARK, BROOKLYN

During our interview with Marianne Monahan at the Long Island home she shares with her three children, the kids were constantly in and out as we sat at the dining room table talking about September 11. Even when one of them caught a pizza box on fire in the kitchen, Marianne didn't miss a beat. She rose quietly from her chair, extinguished the fire, and continued on with her story, demonstrating the unique marriage between her dual roles of professional firefighter and competent mother.

Marianne was a fire lieutenant when the World Trade Center was attacked, and she talked with us about the possibility of being promoted to captain within the next six months. The last time we spoke with her, she had received her promotion to captain and was awaiting her next assignment.

Marianne was among the original 40 women who challenged the Fire Department of New York in 1982 with their groundbreaking entry into what had always been a man's world. She spent most of her career at Engine 298 in South Jamaica, Queens, was a fire marshal for two years, and has been in command of Engine 321 in Brooklyn for the last four years.

Although her personal experiences on September 11 vary greatly from others in this book, Marianne's story drew us in because it so clearly depicts the angst of a devoted, caring, single mom working in a dangerous, unpredictable career that she loves. Her story gives us a glimpse into a world that most people will never see.

"Was it women's intuition? Mother's intuition? I don't know. All I know is that my children have already lost a parent, and I'm all they have."

—Captain Marianne Monahan, November 3, 2001

"As a fire lieutenant, I'm responsible for everything that goes on in my firehouse during the hours I'm there. I'm responsible for making sure all the equipment and personnel are ready for duty. I conduct training drills. I do building inspections and hydrant inspections. My main responsibility, and the one I take most seriously, is directing the members of my company in fire suppression. At a fire, I direct five men at a time, and I tell them exactly what to do.

"I'm in a single-engine company, which means that we respond to areas by ourselves before a truck company comes in. It's my job as lieutenant to go in and locate the fire so I can direct the firefighters about what kind of hoses we're going to use and anything else that relates to fire suppression. There are times when we're the only company on the initial response, and because it's my job to go in and find the fire, I'm often in there by myself.

"Sometimes I'm in there before the fire flashes over, and this is a situation that not many people will ever be in. When I first go into a fire, I can see things very clearly because there's no water on the fire. I might see the mattress on fire, or the fire coming out of the closet, or one time I was at a basement fire and it was just a pile of trash on fire in the corner before the whole ceiling lit up.

"For the last several months, however, I've been working on an administrative assignment for the fire department, so I've been out of the firehouse since April. On September 11, I was in Lower Manhattan, walking to the office from my car. I was looking up at the Twin Towers, and all of a sudden I saw one of them explode. I didn't actually see the plane because I was on the other side of the building, but I heard the explosion and could see the top of the Tower engulfed in flames. I took out my cell phone, called my mother, and said, 'Ma, the World Trade Center just exploded. But don't worry. I'm just going to go down there and see what's going on. Then I'll probably come home.'

"Obviously, at that point in time, I had no idea what I had witnessed. I went into my building and tried to find the other lieutenants who work with me, but I couldn't locate them. By the time I came downstairs, the other Tower had been hit. I couldn't believe what I was seeing.

"Being a woman in civilian clothes, I had on these nice little black sandals that accentuated my outfit, and I realized that I couldn't go down to the site in those shoes. I always carry a pair of sneakers with me, so I put those on and started heading toward the World Trade Center.

"I have 20 years on the job, and most of the men I work with at the firehouse have less than 5 years. Even with all my experience, it's still a problem sometimes being a woman in the fire department. On that day, I figured if I went down there in civilian clothes, they weren't going to be looking at me like, 'Oh, here's one of our officers.' They were going to look at me like, 'What do you want?' I was working on this administrative assignment for Chief Ganci, and I said to myself, *He'll probably be down there. I'll go find him and see what I can do to help out at the Command Post.*

"I was heading right toward the World Trade Center, and hoards of people were running past me the other way. That didn't really bother me because, as a firefighter, I'm used to that. I'm always going in; they're always coming out. I got about two blocks away, and this switch just went off in my head. It was like a light switch. I said to myself, *What are you doing? You've got to go home to the kids.*

"Was it women's intuition? Mother's intuition? I don't know. All I know is that my children have already lost a parent, and I'm all they have. Their father died in September 1999 after being sick with lung disease for about 10 years. He had a double lung transplant and died when he was 42. My sons, Matthew and Mark, are 11 and 13. My daughter, Teresa, is 18. Now we're on our own.

"So there I was, one minute a firefighter bolting through the crowd at a very, very fast pace, and the next minute a mother going home to my kids. I just made an about-face, and without even stopping to think about it, headed back to my car. Of course, had I been on duty that day, I wouldn't have had a choice. I would have claimed my role as a fire officer and gone in with the rest of them without a second thought.

"I know this sounds stupid, but until my husband died, I didn't realize how important I am to my children. It's not something I thought about on a regular basis. I've been criticized by members of my family for continuing in the fire service after the loss of my husband, because now I'm the only parent my children have. I've been criticized for taking chances the way I take chances, although I don't believe you have control over when you die.

"I did the right thing by walking away from the World Trade Center. I know I did. People have said it was amazing that I knew not to go there. If I had, I would have been standing right there with Chief Ganci and all the other people who died. I would have been right there with that whole crowd when the Towers collapsed on top of them.

"By the time I got back to my car, emergency vehicles were screaming into Lower Manhattan. The sirens were deafening. They'd already closed one side of the Brooklyn Bridge, and I was actually the last car they let over the bridge before they closed the other side. People were jumping into my car to get rides. One of the people who got in my car was a man who had missed his train that morning.

Otherwise, he would have been on the 90-something floor of the North Tower, the first one that was hit.

"I had just gotten to the other side of the bridge when I heard that the first Tower had collapsed. When I heard 'collapse,' I thought the top of the building must have fallen off. I wasn't thinking total collapse, as in the whole thing. By the time I got home, my family was in a total state of panic. 'Ma, I'm going down to the World Trade Center,' was the last they'd heard from me. I went over to my mother's house, and they were all hysterical crying at that point. My two boys were in school, but my daughter had been on the phone to my mom about five times. 'Where's my mother? Where's my mother?'

"Prior to September 11, I don't think my children really realized what I did. They knew I was a firefighter, but I didn't come home and talk about exactly what I was doing. After I came home that day, I had to go back for the recall. I know my children were upset that I was going, but I felt that I had to go. Every single firefighter was ordered in. So, I got my gear and went back to my firehouse in Brooklyn. I was gone all night.

"I ended up supervising some fire companies from Jersey. Our battalion had been stripped of apparatus and personnel, so these companies had come over to respond to any kind of fire or event that might happen in our own home area. We had one piece of apparatus left, a big four-wheel-drive truck that we use to go into brush fires, and I had a chauffeur driving me around in that. Even so, I had no mask, I had no radio. My rig was stripped of tools, so basically I was relying on the Jersey firefighters to help me take care of the immediate response area. Fortunately, everything worked out well. We had minor responses, and the guys were great.

"I came home from work the next day, and my children were beside themselves. Once I realized how upset they were, I knew I couldn't go back again. I don't want my children to go through life with this level of anxiety. I don't want them thinking every time I go to work, *Is Mom coming back, or not?* I wouldn't trade my life for the world, but I do have to worry about both sides of it. When I'm 80 years old in my rocking chair, do I want to be able to say, 'I was battalion chief with the FDNY,' or do I want to say, 'I have three beautiful children, they're all happy, and they're all successful'? To me, what's really important is having my kids turn out okay.

"When I called up headquarters to tell them I wasn't coming back to the recall, the officer said, 'Well, I was going to call you and tell you not to go anyway.' As it turned out, all the lieutenants working on this special assignment were pulled off the recall. So it was a wash, even though it was a decision I had already made for myself.

"Since September 11, I've been supervising the firefighters who work at the counseling unit. I see the other side of the devastation, the people who were left behind. I'm not a professional counselor, but I talk to people informally. I talk to widows, mothers, fathers, and firefighters who lost members of their companies. I don't need to go down to Ground Zero. For me, this is enough.

"Sometimes I'm listening to a song and I just start crying. A lot of what I feel is on the personal side, because I know what it means to lose somebody. When they called me and told me that my husband had died, I said, 'I'm coming to the hospital.' I had to go there and actually physically see that he was not alive. Even after the funeral and the wake and the whole nine yards, I was at the cemetery after they buried him, and I was telling everybody, 'You guys go on to the restaurant. I'm going to stay here with him a little while longer.'

"I can't help but think how difficult it must be for the families of the people who died, the ones who haven't been found. How could they not say, 'Well, maybe he's not really there, maybe he didn't go there'? As human beings, we need to have that closure. There's no way around it for these people. It's something they'll have to get past by themselves. I know how difficult it was for me, even having seen that my husband was not alive and having had that closure. This must be really hard for them. When do you have a memorial service? At what point do you say, 'Okay, today's going to be the day we're going to have a memorial service and then we're going to move on'? They have to decide that for themselves. I don't know what that feels like mentally, but I'm sure it has to be a very anguishing place.

"In 1995, I was in a vacant building fire where I lost my lieutenant. I saw him fall through the floor into an inferno. We did what we could, but he passed away. I will never get over that, losing someone at work. It's a horrible feeling. You come back to the firehouse, and it's never the same. It changes everybody. So I can't imagine what it must be like for these firefighters who lost 5 people, 10 people, or 15 people in their firehouses. My heart goes out to them. They're coming in for counseling very slowly, in dribs and drabs, because we're having so many funerals and memorial services. But there was a day about a week ago when there were no services scheduled and a lot of firefighters showed up at the counseling center.

"I think people are still very busy taking care of each other, taking care of the funeral things, and taking care of the services. I don't think they have time yet. The ones I've seen come in look like they've been through war, by the looks on their faces. I remember when I was younger, a couple guys on my block went to Vietnam, and when they came back, they had that look on their faces. They didn't look like the same people anymore. Some of the ones I see coming through the counseling center have that same look. I think that once the funerals are finished, once the memorials are finished, there's going to be a big realization of, 'Now what do we do?'

"I was a teacher before I came on this job. I have a Master's degree in psychology, and I was perfectly happy with my career. To be brutally honest, I was not interested in the fire service at all. It was just an opportunity that presented itself. I had taken the written exam in 1978 because my brother and a friend were studying for it, and I told them that it wasn't a difficult written test and you didn't really have to study for it. They challenged me to take it, which I did. And I passed.

"When none of the women who passed the test were hired, it became a lawsuit, and I was contacted by lawyers to come and train for the physical. The training was over the summer, so I figured, *All right, if nothing else, I'll go get in shape.* Then one thing led to another. I worked out, I trained, I took the physical, and I passed. *Okay,* I thought, *I'll go to the Academy and I'll see how that is.* The Academy was the most physically and mentally challenging place I've ever been in my life, and I felt that I couldn't stop until I had overcome all the obstacles.

"The obstacles were mostly mental. The men. The men hated us. They just absolutely hated us. It was not a secret that they didn't want me there. And the more they opposed me, the more I wanted to be there. If I was going to leave, I wanted to leave having conquered, having been able to say that I was able to do this. I never thought I'd be on the job this long. If you would have told me 20 years ago that I'd still be here, I would have laughed at you.

"I think that having children is what made me get through it. I was only a probationary firefighter when I became pregnant with my daughter and I had to go out on sick-leave because you're not allowed to work when you're pregnant. I had to go to the medical office, which used to be a huge room with pew-like seating where everybody faced the front and you'd walk in the back. If I stayed in the back, nobody would see me. But when they called out my name—of course it's a woman's name—forget it. Everybody would turn around and sneer at me.

"I used to go to the medical office with my brother so people would think I was a wife. When I thought they were going to be near my name, I'd go up to the front so I wouldn't have 200 sneering faces turning around looking at me.

"In retrospect, having my daughter really put everything in perspective. Once you have a child, you realize what's important. I realized that it was a joke that the guys thought they'd get to me because they wouldn't talk to me. At that point, I was able to put the treatment in perspective. Before that, there were many times when I cried all the way to work and I cried all the way home. There were many, many times.

"I was also married to a very traditional man. My husband was born and raised in Italy, and my joining the fire service really put him over the edge. We divorced over it. We subsequently remarried, but I really couldn't come home and tell him what had happened at work. 'Well, I don't know why you don't quit that job,' he'd say. 'You don't belong in that job. What are you doing that job for?'

"What appealed to me about it was that it was it the hardest thing I'd ever had to do. I had always been really successful at whatever I tried to do, any job I held, any school I attended. This was just so challenging because I had to work so hard at it. I had to mentally prepare myself to go to work. I had to be physically prepared to go to work. I had to be on my toes all the time. It was a challenge I couldn't walk away from. If I had walked away from the challenge, I would have regretted it my whole life. I would have been constantly thinking about it. *Oh, I should have done it, I could have done it.* It would have been one of those.

"At this point in the fire service, after 20 years, I pretty much know everybody, so it's kind of funny now when I get a letter from the union that says, 'Dear Brothers.' I'll circle it in red and write, 'I'm not your brother' and mail it back, and they get a charge out of it because they know me.

"It's rare to see a woman in the fire service in this city. There are just 25 of us out of 11,500. When we first went for medical training, part of how we were trained was by practicing palpating, or feeling a patient's body for injuries. So here I am, the only woman with seven guys, and the EMS instructor says to me, 'Why don't you lie down and we'll practice on you?' And I'm like, 'Are you nuts? There's no way I'm lying on this floor and letting these seven guys put their hands on me.' It wasn't a big thing, but don't you know, they made a big thing about it, and I had to get involved in setting the whole protocol about how female firefighters should be handled at EMS training. It really was just common sense.

"I would like people to know that every woman in the FDNY is totally dedicated. That's not to say that the men are not, but every single woman has given of herself 150 percent. I find it very difficult that we're still looked upon as such a novelty. There's not a day in my career when someone in the public doesn't have a comment to make like, 'Oh, I didn't know there were women in the fire service,' or 'Oh, look at that! Oh, are you a girl?' Not a day goes by when that doesn't happen to me, and after 20 years, let me tell you, it's old. I'm dedicated, just like my sister firefighters are dedicated to their jobs, and it's very frustrating at this point in my life, in my career, that there's still that little 'ha ha ha' thing going on in the public.

"I realize it's not the public's fault, but it also makes me question the perception of society as a whole about women. I'm a firefighter and I'm a mom. I'm going into burning structures and I'm up at school at the PTA. We can be both. I *am* both.

"Just as a man would love his job as a firefighter, a woman loves her job as a firefighter, and it's nothing less. This is the job I've chosen to do, just like every other woman in the FDNY, and it's not something that any of us takes lightly. The passion I have for it, the willingness to learn, and the knowledge I've gained is certainly no less because I'm a woman."

410
SCUE
410

BONNIE GIEBFRIED

AGE 37
EMERGENCY MEDICAL TECHNICIAN
FLUSHING HOSPITAL MEDICAL CENTER
FLUSHING, QUEENS

We met up with Bonnie Giebfried on Long Island, where she's a volunteer firefighter/EMT with her hometown fire department. Bonnie all but grew up in the firehouse, where her father and grandfather are both former chiefs. Her family has a long history of volunteerism in the fire service.

On September 11, Bonnie was working on a 911 ambulance out of Flushing Hospital Medical Center in Queens. Although she and her partner had little knowledge of Lower Manhattan, they didn't hesitate to rush to the scene of a plane crash when they were dispatched to the call.

"At the time of the dispatch, only one Tower was burning," she told us. "Even though I've been a firefighter for five years, I'd never seen a high-rise fire. We had no idea what we were getting ourselves into."

As a kid, Bonnie watched the Twin Towers being built.

"It was exciting watching the Towers go up," she said, "because it was big news. They kept getting higher and higher, and everyone said, 'They can't go any higher than that.' But they kept going up and up."

As Bonnie and her partner were treating patients in the shadow of the South Tower, she said it never crossed her mind that it was going to fall.

"I don't think that anybody who was standing down there, looking up and watching the debris coming down and the people falling, thought the buildings were ever, ever going to fall," she said. "I still can't believe they're gone."

"Our breathing was getting worse and worse. We were suffocating. We were literally starting to die, because the air was depleting. No light, nothing. All we could think was, *Get out. Get out.*"

—Bonnie Giebfried, January 21, 2002

"I'm an emergency medical technician, and I work for Flushing Hospital Medical Center in Flushing, Queens. We're a voluntary ambulance that works through the 911 system, and we're subcontracted through the Fire Department of New York. We're trained the same way as the FDNY personnel and we've got the same certification, but we're not part of the fire department. My paycheck comes through the hospital. We don't have the illustrious benefits that the fire department people do.

"We ride in pairs, and we never know what's coming. Most of my partners are long-time veterans of the system, but I've only been here about a year, so I'm kind of the new kid on the block. I'm also a volunteer firefighter/EMT with the Oceanside Fire Department, which is in my hometown. I'm part of Rescue Company One, which is a search-and-rescue, heavy rescue, and ambulance corps. We have six houses in Oceanside, which include two hose companies, two pump companies, one ladder company, and one heavy rescue/ambulance company—that's me. I'm a volunteer, and I'm working on my fifth year now.

"I think this kind of work has always been in my blood. I've been involved with first aid since I was a kid in Brownies, then I was a lifeguard, and I went from there into the health-care profession. I was a social worker, so I had to have CPR certification and first aid training. It seemed to always follow me through my years.

"September 11 started out like any other day in Flushing, Queens. My partner, Jennifer Beckham, and I started work at 5:30 A.M., and after we checked our ambulance, we got our breakfast and went to our post on Northern Boulevard to await our first call. We were eating our bagels and talking, and we were listening to a morning radio show. They're always joking around and pulling gags on the morning show, and they said something about a plane hitting the World Trade Center. I said, 'You know, Jennifer, that's not even funny.' She said, 'No, I think it's really true.' She turned the volume up, and they said it again.

"Then our radios started getting busy. The dispatcher came on and said, 'I need the following units for an MCI,' which is a Mass Casualty Incident. She started reading off a list of ambulances, and it seemed like it went on forever. I said, 'How big can this plane be?'

"Then we got assigned. Jen's from Suffolk County and I'm from Nassau County, and neither one of us knows the city. We got on the Long Island Expressway, and all of a sudden traffic came to a dead stop. People were out of their cars, just standing there, staring at the smoke over Manhattan. When we got to the tollbooth, they'd closed down the whole tunnel to civilian traffic.

"The second airplane hit when we came out of the tunnel. At that point, I think it dawned on us that we were in way over our heads. But we kept on going. As an EMT or firefighter, we just override the fear. We have a job to do, that's what we're dispatched to do, and that's what we're going to do. We just went on autopilot.

"By the time we got downtown, it was really congested. The first place I wanted to park was under the pedestrian crossway to the Financial Center, because I figured it would give us cover. But they kept flagging me on, and there were droves of people in the streets. We parked where we were directed and noticed the stuff coming out of the Tower we were under—papers and ash and smoke. Then we started seeing all the bodies, the parts of bodies. This was way before anything collapsed. We were seeing the aftermath of the second explosion.

"We got our stethoscopes and IDs and radios out of the cab, then we went around to the back and got our helmets. We got the stretcher, the backboard, the cervical collars, the medical bag, the defibrillator, the suction—we pulled everything out of the bus and put it on the stretcher. Did I realize what we were in for? No. Did my partner know? We had no clue.

"We met up with EMS Captain Karen DeShore. She said, 'We have three people who need help in the South Tower, and I want you to go in.' We said, 'Okay,' and we grabbed our gear and started to maneuver toward the building. We had our helmets on, but we weren't really protected. We didn't have bunker gear, and we didn't have masks. Across the street there were all kinds of debris and things were on fire. As I think back now, any one of those cars could have blown up and we would have gone with it.

"We got to the Tower and found our three patients. One woman was paralyzed with multiple sclerosis, and they just wanted to get her to safety. I picked up the woman and put her on the stretcher. The other two women were really scared, but they weren't hurt. I said, 'Listen, stick close to us, keep your head ducked, and when we say, 'Go,' don't stop anywhere; just keep on going. If you get scared, just run forward. Don't go back.' I gave the older woman my helmet, and Jen gave the other woman her helmet, and across the street we all went.

"When we got across, someone went back for the disabled woman's motorized wheelchair, and we told the three of them, 'Go toward the river. There's safety at the river.' We were regrouping at our ambulance on the southwest corner of the South Tower, which was only about 500 feet away. I had to strain my neck to look up to see everything that was going on above me.

"Captain DeShore told us that the debris was coming down too close to us, and she moved us back about 12 or 15 feet onto the grassy knoll of the Financial Center. That was our saving grace. All of a sudden, the Tower was blowing up and imploding, and a fireball was coming at us. We weren't even a full block away.

"I grabbed Jen by her shoulders, threw her in front of me, and we started running up the grass. We went around to the side of the building and ran into an alcove, because we thought the doors were there. Well, there were no doors there. We heard things banging against the building, and then everything went dark and we got buried alive. We were buried under trees and cars and everything.

"I guess the trajectory of the debris made it hit the building first, then it slid down and piled up over us. The alcove created a small void, and we were safe between the building and the pile of debris. But we were trapped. We tried to turn around, but there was a wall of debris behind us. There were quite a few people with us, and everybody started trying to break the windows to get into the building. We were hitting them with our fists and our radios, whatever we had in our hands. But we didn't have much luck—you know, windows in the city are three, sometimes four, layers thick.

"Our breathing was getting worse and worse. We were suffocating. We were literally starting to die, because the air was depleting. No light, nothing. It was completely, completely black. All we could think was, *Get out. Get out.* Everybody was screaming 'Smash the window! Smash the window!' Then it got really quiet, and I think that's when it dawned on us that we might not get out of there. I could barely hear people inhaling and exhaling. I could hear my heart beating.

"I said, 'God, take care of my family and my loved ones. Take care of my friends.' Then Jennifer said exactly what I was thinking. 'This is a real sucky way to die.' Standing up, buried, not being able to get free. Some of us were up to our hips in debris, and there was a wall of debris behind us. We were buried pretty deep.

"Everyone was about to stop breathing when I heard *Pop pop pop crash!* The police officer who was in front finally got to his pistol, put it against the window, and fired three shots. It shattered the glass a little, then everybody started banging the crap out of the windows to break them open. When we broke through the glass, people helped each other get through, and we finally got inside the building. All of us got out. Every single one of us got out.

"We started throwing up, because we had to get whatever we had inhaled out of our lungs. We went out the back of the building and came around the other side. Everybody's eyes were burning, and ash and debris were still coming down. People were coming out of the rubble. It was like a graveyard. People were just sitting up, coming out of the dust, and asking for help.

"We went into a deli, and Jen and I took some water and washed each other's eyes out. Everybody's eyes were inflamed. People started coming in with cuts, bruises, and broken bones. We still had one of our medical bags with us, so we went into our EMT role and started treating people in the deli. There was fresh water that we used to clean people's eyes out, and we told people to swallow orange juice quickly so they would throw up. I just kept on chugging orange juice until I started throwing up again, because I had to get all that crap out of my system.

"So I puked and treated people, puked and treated people. Head injuries, arms, shoulders. At one point, a guy said there was fresh air in the freezer. It was our only supply of fresh air, so we went in and out of the freezer between doing everything else. We handed out slings and cravats for people to put over their faces, because we didn't have any masks to give them.

"Then Jen and I looked at each other and she said, 'Hey, if that building fell, the integrity of this building may be compromised.' I said, 'Let's get the hell out of here.' We came out of the deli and walked over to the fire department's Mobile Emergency Response Vehicle to get some supplies—bandage bags, water, just basic things we could use to treat minor injuries. As we were walking away, I felt the ground shake again, and the North Tower started to come down.

"That cloud of debris was coming at us again, and we ran down into a parking garage. Just by instinct, maybe because I went through Naval Prep School, I threw Jen down and jumped on top of her. I figured that if something was going to come at us, I was a bigger blockade, so I grabbed her and tucked her underneath me and said, 'Keep your head down.' There was so much crap coming in.

"Once the debris settled a bit, we got up. Jen had a flashlight, and she flashed it through the debris. I literally could not see beyond six inches in front of my face. We could hear other people in there. One guy said, 'How many of us are here? Someone went, 'One,' and then people numbered off, 'Two, three,' and it stopped at seven. We knew we had seven people with us. We started with seven and we were leaving with seven. I said, 'Come this way. I know that the wall we're closest to goes to the outside, and we gotta go up this incline.'

“We all started kicking the debris away, pushing it away from the door of the garage. If it was heavy, we didn’t notice. I think our adrenaline kicked in, and we pushed our way out. We saw another EMT coming out of the rubble, and I grabbed hold of her and said, ‘Listen, just follow us. You’ll be okay.’ I took her hand and calmed her down. We went through the courtyard and down the stairs, and all of a sudden everything opened up and we could see the sun again. We were at the river.

“We set up again, and we started treating people. When we came out of the garage, there were medical bags all over the place. We kept picking up bags. God knows how many bags we carried down to the river. When we got down there, we just dropped all the bags, flipped them open, and started washing out people’s eyes.

“We had two civilians helping us, and one was a nurse. Our paramedic student found us, and then a trauma nurse came and started working with us. We all took handfuls of triangular bandages and stuck them in our pockets and started handing them out. When we saw people who were hurt, we’d sit them down and start triaging. We helped people with the little minor things, and if it was a major injury, we told them to sit on the bench and just wait there to be transported to a hospital. As all of this was going on, the fireboat came in, and fishing boats were picking up people and taking them off.

“We put the worst patients on the fireboat. If they weren’t so bad, we fixed them up and said, ‘Just keep on walking south to the marina. Somebody will help you get on a ferry.’ We kept on doing this until I started getting sick. The debris that I inhaled was precipitating an asthma attack. I hadn’t had an asthma attack in years. We were out of oxygen, so Jen went into an FDNY ambulance to get Albuterol treatments and oxygen for me. I had to take a treatment right there. I was wheezing, and my airways were really compromised.

“I didn’t have time to be afraid. It didn’t occur to me to leave. Jen said, ‘Get on the boat! Get on the boat!’ I said, ‘I am not leaving this island unless you’re coming with me.’ Someone got me some water, and I threw everything up. My airways opened again, and I said, ‘Listen, I feel better. If I start getting wheezy again, then we’ll think about getting off the island.’

“Eventually, Jen acquired one of those little sanitation golf carts that picks up garbage. We threw all our stuff on there and headed for the marina. We picked up firefighters and a fire chief as we went, so there were about eight of us on this little sanitation golf cart. There we were, racing over all these hoses and past all the firefighters and ducking underneath their ladders.

"We got to the seawall where the marina was, and someone screamed, 'There's going to be a gas explosion!' At that point, I got very dizzy, and I was afraid I was going to fall down. I said to Jen, 'It's going to be a bad attack. I feel it coming.'

"We got onto the police boat, and I had a heaviness in my chest and a hard time breathing. I remember Jen saying, 'Keep on breathing. There's no oxygen or anything on the boat.' I said, 'Okay.' I looked to the right and saw the Statue of Liberty. I looked to the left and saw the smoldering of the city. I don't know if I blacked out at that point, but I don't remember anything else until the boat started slowing down.

"I tried to stand up, but I couldn't. Two guys picked me up and got me on the dock. I threw up again, and I kind of went limp, and they just picked me up by my pants and under my shoulders and handed me off to a paramedic. I had bad chest pains, and I thought I was going to have a heart attack.

"I said to Jen, 'I'm going to crap out. I feel it coming.' She got the paramedic to come over, and there was a problem because I'm allergic to so many medications. They finally started an IV and hit me up with a medication that started to break up the chest pain, and the next thing I knew I was being whisked off to a hospital in New Jersey.

"Both Jen and I were treated at the hospital. They put us in a treatment room where the news was on, and all we saw were the Towers blowing up over and over and over again. I said to Jen, 'Wow, did you see that fireball? I guess that's what came at us.'

"It started to really sink in when I saw the top of our ambulance, 4852, on TV. It was buried in debris. It was destroyed. I said to Jen, 'That's where we were standing!' If it hadn't been for Captain DeShore moving us back just before the Tower collapsed, we would have been dead. She saved our lives.

"Then we started hearing the Army helicopters coming in, and then the F16s were coming in, and the hospital was getting busier and busier. It was all starting to become real to me.

"They were going to let Jen go home, but they didn't want to let me go because I wasn't stabilized. They kept pumping medication into me. I said, 'I'm not staying in New Jersey, because it's too far away from my family.' I said, 'If she's going home, I'm going home.' Jamaica Hospital Medical Center is our mother hospital, and I wanted to go to Jamaica, because that's where my home base is. So, they sent a van from Jamaica Hospital to come pick us both up.

"When I got to Jamaica Hospital, I had my third asthma attack. I spent two days in the hospital, staring at this billboard of the *Titanic* sinking outside my window. I broke a few times in the hospital. I was crying because I was scared. I was on the fourth floor, and I didn't want to be up in the air. I was in this private room, and they were leaving me alone, which was the worst thing they could do. I was staring at the walls, and the psychiatrist said, 'No TV.' So no TV, nobody was interacting with me, I didn't have a phone, and I was really cut off from the world.

"Everybody else went home that night, but not me. Psychologically, I think I'm behind, recovery-wise, because the physical stuff has really hindered me. I still don't feel well. I still get chest pains, and the headaches haven't gone away. My eyes go blurry. I haven't had any more asthma attacks, but the wheezing and the breathing are still a problem. I'm on more medications than you can imagine. I go to work and take my medication. Go to work, take my medication. I take care of my grandmother. I see my friends, and my friends and I are very tight. My family has become overprotective.

"The lack of recognition for everybody who was down there is probably the thing that makes me the angriest. We had civilians who came off the streets to help us, and they'll never get the recognition. The fire department and the police department get plenty of recognition, but EMS is the stepchild of the fire department and hasn't really been recognized for being down there.

"The media clumps EMS under the fire department, and we're really separate from them. That's the biggest, most frustrating factor right now. Everybody's under the impression that we're the fire department, and we're not. They think we're being taken care of, and we're not. The people who are in the voluntary ambulance corps, who work for hospitals and private companies, are not being taken care of in the same way as the people who work for the FDNY, either physically or mentally. The majority of us are taking money out of our own pockets to go to the doctors.

"The only assistance we got was from the Red Cross. The United Way didn't want to have anything to do with us, because we were workers who were still getting paid, and they said we didn't need assistance because we had our own coverage. That's a myth. A big myth. In my case, if it weren't for the Red Cross, I would probably be buried under a good $20 grand in debt right now. I was off work for 48 days, and I can't afford that. I live with my grandmother who's 88 years old. She's on a fixed income, and I'm supporting the household. I'm the main caregiver for her.

"I have daymares and nightmares. I see the fireball coming at us and then the darkness. I was afraid of the dark for the longest, longest time. I could not go to sleep without the lights on for about the first two and a half months. Then it was a small light, then it went to a night-light, and now it's the TV that shuts itself off after I fall asleep. As long as I'm not sitting in the dark thinking about what happened, I'm okay. I've awakened from nightmares where I thought the world ended and I'm the only one left, sitting in my bed.

"Part of me died down there. It's gone. Not coming back. I'll never, never be the same. I'm keeping myself very busy, so the impact is not really hitting me. But I'm still feeling the effects of it. I've written more since all this, and I've taken a lot more pictures. I'm Miss Nature Buff, so nature is my comfort right now. It's the simple pleasures that help me get through the day—my family, the ones I love, the things I didn't take time for before this.

"My mother and I went to the beach last night. It was freezing, and the sunset was so pretty. Then it got cloudy, and I realized, if you don't have one, you don't have the other. Everything is about trying to find that balance, and the balance for me right now seems to be in nature. Nature, no matter how badly we screw up the environment, always comes back. I go outside and look up and hear the geese and say, 'No matter how much we've screwed things up, these creatures persist.' Nature persists. The sun will rise and it will set, no matter what we do."

1370
NYPD
POLICE DEPARTMENT
NYPD 9-11
1370
DONAHOO

TRACY DONAHOO

AGE 33
POLICE OFFICER
NEW YORK CITY POLICE DEPARTMENT
TRANSIT DIVISION, DISTRICT 2
LOWER MANHATTAN

Tracy Donahoo is a rookie police officer who graduated from the Police Academy just a few months before the World Trade Center attack. Although she had the least amount of experience of anyone we interviewed for this book, her story proved to us that courage is not necessarily commensurate with the number of years on the job.

Tracy used her instincts to save herself and the people she was helping to evacuate from Five World Trade when the first Tower collapsed. A compassion for others and a deep desire to help kept her focused on the job she had to do.

It took a while for Tracy to decide she wanted to be a police officer. She told us that she'd been a waitress and a bartender, had several desk jobs, and worked in the garment industry for eight years. She has always been searching for something more meaningful.

"At the end of the day, if a dress wasn't made, it really didn't tear me up," she said. 'It wasn't earth-shattering, the way a lot of people perceived it. I wanted to do something important to help people, like being a nurse, a firefighter, or a cop. I suppose we should all be dressed properly, but you can wear something from K-Mart as well as something from Saks. It doesn't really matter in the end."

"There were steps in the middle of the store, and people were tripping as they were coming through, so I stayed behind with my flashlight to get them safely down the steps. I was the last person to walk out of there."

–Tracy Donahoo, November 24, 2001

"September 11 was my second day on patrol. I graduated from the academy in May, then did three months of training, so this was only my second day as a real police officer.

"I was assigned to the Broadway Nassau subway station at Fulton Street and Broadway, which is a big complex on many different levels. My partner, Carol Paukner, and I had just completed inspecting the station, which is one block away from the World Trade Center, when we heard a 1013 over the radio. A 1013 means to leave the post where you are and respond to a priority call.

"I ran up the stairs and ran down the block, and as soon as I hit the street, I was like, 'Oh, my God!' There were a lot of vehicles blocking the street, because everyone had popped out of their cars. They were standing underneath the North Tower, looking up at the fire, dumbstruck. We started moving the vehicles out of the way so emergency vehicles could get through.

"Then I went up to Five World Trade. People were going down to a sublevel from the North Tower and coming up through Building Five, and we were trying to get them across the street. They were taking it very slow. They were saying, 'Oh, I just walked down 70 flights. Don't make me walk any faster.' We were trying to keep them under the roof, because there was still debris coming down from where the first plane hit. People were actually lined up over by Trinity Church, just standing there, staring at the fire.

"One of the officers behind me said, 'Oh my God, Tracy, another airplane is coming!' I could hear the plane just coming and coming, and the engine was getting louder and louder. Then I heard it hit the South Tower. There was a shower of debris and parts of the plane. I took cover with a couple other officers underneath the lip of Building Five, near Borders Bookstore. Airplane parts were falling and crushing police cars, and people were screaming. We were slamming into one another trying to break into the bookstore through the revolving door, but the store wasn't open yet and we couldn't get through.

"We went back to evacuating people, and at this point, it wasn't a calm walk across the street. It was more like, 'Run!' We didn't know what part of what was coming down. I remember a blind man coming out with his dog, and I was yelling to someone, 'Grab the dog!' I know you're not supposed to take the dog, but we had to get him across the street quickly and safely.

"We did that for a while, and then there was a sound. I turned around and looked into the windows of the Millennium Hotel, which was across the street from the World Trade Center, and I could see the reflection of a building crumbling from the top. It started waving and then it just cascaded down. That was the South Tower coming down.

"I turned around to the people who were still coming out, and I yelled, 'Get back in the building! Get back in the building!' I thought it was kind of an avalanche. You could see the fear in people's faces, because here's this police officer screaming, 'Get back in the building!' and I was running toward them. Just when I got into the building myself, the South Tower came all the way down.

"The force of all the debris hitting the ground sent me flying. I must have gone 20 feet. When it hit, it was the loudest sound I ever heard. I went soaring through the air, and I thought, *Oh my God, this building is going to land on me.* I was just waiting for it—the pressure, the pain.

"My mouth, ears, and eyes filled with debris. I had to stick my hand in my mouth to pull out the debris, because I couldn't breathe. Then I was so angry with myself. I said to myself, *You dummy. Why did you run into this building? This is just like Oklahoma City. You're going to be trapped in here, and they're not going to be able to rescue you. You're sealed in here now.*

"I was in a vestibule, like a lobby area, and it was pitch black. I couldn't see a thing. In the Police Academy they say, 'Never give up.' So I thought, *I'm not dying in this building.* I got up and pulled out my baton and my flashlight, and everyone was screaming and screaming. I said, 'Shut up! Shut up! Shut up! I'm a police officer!' And then they were quiet. I could see all the flashlights go on, so there were obviously other people in there—firefighters, police officers, and I don't even know who else, because I couldn't see anyone. All I could see was the shimmer of their lights.

"I said, 'We're going to get out of here!' I hit my baton forward because I wanted to know how far I had flown. I knew that building very well. I'd been in there a million times. I hit a solid wall, and I thought, *Okay, I'm by the MTA police station.* I turned my flashlight on the woman behind me, and she said, 'Don't leave me!' I said, 'I'm not leaving you.' I shone the light around, and there was a firefighter next to me. I couldn't really see him, except for the shimmer of the reflective strips on his jacket and the hook, whatever that hook thing is that firefighters use to break walls.

"He said to me, 'I have to get out of here. My buddies are trapped upstairs.' I said, 'Well, there's a bookstore in the lobby, and it's glass all the way around. There has to be a way out.' I moved my flashlight toward the bookstore, and then everyone started going in that direction. We walked into the bookstore, and it was perfectly intact. It was amazing.

"The firefighter walked ahead of us and broke the glass. There were steps in the middle of the store, and people were tripping as they were coming through, so I stayed behind with my flashlight to get them safely down the steps. One lady walked by and said, 'Thank you,' and I said very calmly, 'Oh, no problem.' I guess I didn't really understand what had happened. I was still kind of stunned. I was the last person to walk out of there.

"I ran into a female sergeant from the Port Authority Police Department at the door just before I went out, and she was shaking. She said, 'I lost my partner.' I said, 'I lost mine, too, but I'm sure they're okay. Let's walk together, and I'm sure we'll find them.' So we linked arms and started walking. We walked outside, thinking we were going to walk out into sunlight, but everything was covered with white debris like a snowfall. We could only see a little way ahead of us.

"We ran into a couple of other Port Authority police officers, and she stayed with them. I continued on to look for my partner. I ran into a woman walking along the street who only spoke Spanish, and she was hysterical. I could hear someone yelling, 'There's first aid over here! There's first aid over here!' So I walked her over there, and I put water into my eyes to try to get some of the debris out.

"It was just chaotic. I figured I was going to be there all day, and I needed to take a bathroom break. I asked where a bathroom was, and when I got there, it was a men's bathroom. There was a man inside, but he said, 'I'll be right out.' Then he came out and I went in, and while I was in there, he started screaming that the building I was in was going to collapse. He was yelling, 'The building's going to collapse! Get out!' I ran out and had my pants up, but I didn't have my two belts done, so as I was running I was yanking on my belts, trying to get them on. It turned out that nothing happened to that building.

"I walked down one block to Broadway, and I ran into a sergeant from my command. He said, 'Oh my God, you've got to call the command. They've been calling your shield number for an hour.' I got to the payphones, and there were a bunch of reporters on the phones. I said, 'I have to use the phone,' and they said,

'In a minute.' I said, 'No, you have to get off the phone. I have to call my command.' Another reporter standing there said, 'No, I'm using the phone next.' Then he realized I was a police officer, and he said, 'Oh, sorry, officer.' I was surprised he could even tell I was a police officer, because I was completely white. I had all this debris on me, and my hair was totally saturated with the stuff.

"I was on Broadway, across from City Hall, on the phone with an officer from my command when the second building came down. I heard a *Bang!* and then this black smoke came at me so fast. I saw ambulances flying down the street and people running and screaming, and I dropped the phone and ran to the next corner and jumped behind a building. All the smoke and debris went past me.

"About half an hour later, I finally got in touch with my command again, and the officer I'd been talking to was so upset. He said, 'Oh my God. I thought you were dead. I thought you were killed when you were on the phone with me.'

"All I had was a little cut on my ear. I had some skin burns, and they didn't know when I went to the hospital whether it was from asbestos or cement or whatever. My corneas were scratched, and I was a little black and blue from falling on the ground, but my injuries were really nothing. I only took one day off, and I felt fine. None of the officers who worked at the 1st and 5th Precincts and Transit District 2, which are the units that handle all of Manhattan South, got killed. I can't believe that no one in my command died. That's our place. Our trains are right there, right underneath it.

"I think I was calm during the whole thing because I was stunned. I didn't realize what a terrible situation I was in. I wasn't hysterical. Even when I went back to my command, I told the story over and over again and I was relatively calm.

"People ask me if I've cried. A month later, I took my mother down to the site, and that was when I cried. But other than that, I haven't. Maybe it's because we've been so involved on a daily basis. When I look at the buildings, I'm just in awe. The destruction is unbelievable.

"The Borders Bookstore is perfectly intact. One entire floor has all glass and curtains, and there's some broken glass in one section, but there's a live tree still in there, which is amazing. The rest of the building is totally gutted.

"Two weeks ago, we had another plane crash in Rockaway. I couldn't believe it. My brother's a lieutenant, and he called me up and said, 'Tracy, a plane just crashed in Rockaway.' All my family lives in Rockaway, and I was like, 'Oh my God. Was it on purpose?' In my lifetime, I never thought I'd ask if a plane crash was on purpose. He said, 'They don't know.' I said, 'Are we being mobilized?' And he said, 'I don't know yet.' I could feel myself starting to cry. I said, 'I can't believe this could happen to us again.'

"That plane crash ended up being an accident. The plane landed on four houses, killing those people, but it could have been so much worse. I live about 15 minutes away from there. Rockaway is a beach community. It's about four major blocks across, and there's water on both sides. You're either on the bay or the ocean. It could have just wiped everything out. I went in and took a shower, and I was ready to go. All I wanted to do was help.

"Before this, I wasn't sure I wanted to be a cop. The process of deciding whether or not I wanted to be one was a long one. But during and after September 11, I knew that this is what I'm supposed to be doing. I've always been the type of person who likes to help people, and until now, I'd never really found my niche. The helping aspect of being a police officer is what I like about it. I'm glad I was there. We saved thousands of people. Thousands of people got out of those buildings.

"I want a better world for everybody. There are a lot of children who are missing their parents now. We're coming up to the holidays, and this is horrible. There's a kid on my street whose mother is gone now. I'm angry that so many people died.

"I always expected a physical confrontation or a shoot-out to be my worst nightmare as a police officer. I never thought about something like this. It was only my second day on the job, and I think it can only get better from here."

204 ENGINE CO. 204
PROB
ENG 204
3606

TRACY LEWIS

AGE 29
FIREFIGHTER
FIRE DEPARTMENT OF NEW YORK
ENGINE 204
BROOKLYN

Tracy Lewis had been a probationary firefighter for less than a year when she faced the biggest challenge of her career on September 11. Although she was new on the job, she didn't let the overwhelming events of that day change her mind about wanting to be a firefighter. In the months that followed, she dedicated herself to finding her fallen comrades, even though the work at Ground Zero made her physically ill.

We interviewed Tracy at her home in Brooklyn, where she takes care of her grandmother in the same apartment where she grew up. Paramedic textbooks were spread across the dining room table, and a 10-speed bike was parked next to a hutch filled with a lifetime of collectibles.

After our interview, Tracy took us to Engine 204, the oldest working firehouse in Brooklyn, to photograph her for this book. The photo session attracted quite an audience of neighbors, one of whom was a teacher at a nearby grammar school.

"I didn't know there was a woman firefighter at this house," he said, handing Tracy his card. "Will you come and speak to my class and tell us about your experiences?"

Tracy hesitated at first, but then agreed. "Part of my responsibility as a female firefighter," she said, "is to let little girls know that they can do this, too."

"We dug up crushed fire extinguishers and mangled air packs. Every time someone would find a void, it brought hope, like maybe somebody's under there."

—Tracy Lewis, January 27, 2002

"I worked as an emergency medical technician for the FDNY for a little more than two years. I liked the excitement of being an EMT. I liked the rush. I'd come to work, jump on the ambulance, and we'd be out there listening up for whatever came in. If we heard that there was a bad car accident or that someone fell from a scaffolding and broke his leg, we rushed to get there. I liked knowing that I could do something to help.

"After two years, I had an opportunity to take the firefighter exam. Being a firefighter isn't that hard, but it's different than working on an ambulance. Every day it's something different. One day we might have a little garbage fire, and the next day we have a whole basement on fire. It's kind of weird being happy to go to a fire, but it's the adrenaline rush I like. It's the excitement. That's what it is for me.

"I've been at this firehouse in Brooklyn for about a year, and I'm also taking a paramedic course. That's just something I'm doing on my own so I can keep up my skills. It's something I'm doing for myself.

"I was at home on September 11. I was on the phone with one of the guys from my firehouse, and he was at home, watching TV. All of a sudden he said, 'Oh my goodness, turn on the TV! A plane just crashed into the World Trade Center!' So I turned on the TV and saw the Tower in flames, and some guy on his cell phone was giving a description of the scene. This guy was saying, 'I don't believe it, I don't believe it,' and then he started screaming, 'Oh my gosh, there's another one, there's another one!' and then everything went blank. The TV just went out.

"I hung up the phone and tried to call the firehouse, but I couldn't get through. Over the radio, they were saying that all emergency service workers were to report to their stations, so I called a friend of mine and he took me in to work. On our way, all we could see were ashes, little bits of ashes, and the expressway was empty. There was only one cop on a motorcycle. It was really eerie, because there was total silence.

"Our firehouse isn't far from the Brooklyn Bridge, and when we pulled up, I could see the smoke coming from Lower Manhattan. This was a little after 10 A.M., and as I was getting out of the car, I heard the radio announcer say that one of the Towers had collapsed. I said to myself, *Okay, right. Maybe part of the building fell, but the Tower didn't collapse.* I got out of the car, got my gear out of the trunk, and went into the firehouse.

"It wasn't long before the second Tower came down, and I knew this was for real. Engine 204 is a single-engine house, and they'd already sent it into Manhattan. The captain fit all of us and all our gear into his car, and we met up with hundreds of other firefighters at Engine 211 in Brooklyn. From there, we loaded up on buses and went downtown.

"It was like a parade of people walking across the Brooklyn Bridge. Crowds and crowds of people. I was looking for my brother, who was in the city, because I hadn't heard from him, and I was also looking for my cousins. We actually sat there for a while in the buses because there were so many people on the bridge that we couldn't get through.

"When we got to the other side, it was a ghost town. There were very thick ashes everywhere, almost like a blanket of snow. Before we got off the bus, our captain told us, 'I don't know what we're going to see. You may see friends. The condition you see them in may not be the best of conditions, but I'm going to need you to keep your heads on.' I guess he was trying to prepare us for the worst.

"We got out of the bus, and there was just total silence. I didn't hear anything. I said to myself, *I gotta be dreaming. This is definitely a dream.* It didn't seem real. There were no people, and it seemed like we were in the middle of nowhere. I knew we were in Lower Manhattan, but there was nothing there. Just ashes all over the cars in the street. There was total silence.

"I was looking around and asking, 'Where were the Twin Towers?' I couldn't even tell where they'd stood. There was this one big piece from the building frame that stuck right up in the middle of the rubble. It was like something from a movie.

"Then the captain said, 'Okay. We're going to go this way.' As we were walking, there was a drugstore, and he said, 'Go in, see if you can get a flashlight, and get yourself a bottle of water. We don't know how long we're going to be here.' So we got some lights and we got some water. That's when I saw Ella McNair, one of our senior women firefighters, and I said, 'Oh, thank goodness, you're okay!'

"We started walking again, and we were asking people, 'Did you see Engine 204?' Everyone we asked said, 'No, we didn't see 204. We didn't see 204 at all.' I was just praying and hoping for the best, but the guys at this point were thinking the worst. I was saying, 'Listen, just because nobody's seen Engine 204 doesn't mean the guys aren't okay. I'm sure the guys are fine and they're just waiting for us somewhere.'

"As we were walking, we'd see firefighters and rescue workers sitting on the ground. Some of them were hurt, and some of them were just sitting there in a daze. We continued asking people, 'Did you see Engine 204?' Finally, the captain asked one of his friends, and he said, 'Yeah, I saw 204, but the rig is crushed. I haven't seen any of your guys.' Everybody was thinking the worst, but the captain said, 'Let's just keep looking. We're going to find them.'

"We were walking through thousands of people at that point. We walked this direction, that direction, every direction, looking for this chief we were supposed to report to and at the same time asking, 'Has anyone seen 204?' We'd see a restaurant and the windows were broken, there was no front glass, and debris was scattered everywhere inside. And we'd see people outside covered in ashes and dust and dirt. It still seemed to me like we were shooting a movie. This is something you see on TV. You don't see this when you walk out of your house.

"We ran into one company in our battalion, Engine 279, and the guys were just sitting there with their heads down. I didn't know it yet, but they'd just lost their whole engine company. It wasn't registering. It wasn't registering with me yet that we had lost people. I was just thinking that people got hurt. I was still saying that my guys were fine.

"When we got to West Street, we saw rigs that were crushed, a battalion chief's car that was burned, and a police car that was burned. I saw an ambulance I recognized from when I worked in EMS, and I was thinking, *Gosh, I hope those people are okay.* I was thinking that if the ambulance was here, they couldn't be too far from it. And if they weren't too far from it, then where were they? We saw where a rig had parked under an elevated walkway that went from one building to another, and the whole walkway had collapsed on the rig. I was thinking, *Okay, where are these guys? Where are all these people?*

"We must have walked for hours that day. We saw all these volunteer ambulances coming in from Boston, New Jersey, and Pennsylvania. There was a whole brigade of ambulances, one right after another. We saw other rigs from our battalion, and parts of them were bent or crushed. The glass was broken, things were hanging off the sides, and the lights were broken. We looked everywhere for survivors.

"Then we saw one of our guys, and he started screaming to us. Three or four of our guys were sitting on a corner, and they were just as happy to see us as we were to see them. Two of our guys were injured and had gone by ambulance to the hospital, but everyone was alive. One guy broke down, and we gave them our cell phones to call their families and let them know they were okay. It was a relief to see them. It was really a relief.

"We were assigned to the bucket brigade, which was a line of people passing buckets. We happened to be at the end of the line, and we were the ones who dumped the debris onto the truck. After we'd done that for a couple hours, we hooked up lines to put little fires out here and there.

"The scariest thing I saw that day was when Seven World Trade came down. We were standing on a corner looking up at this building, and it was 47 floors of fire. We were saying, 'We've just got to let that burn. We can't put it out.' We could see people in nearby high-rises looking out their windows. All of a sudden, we felt a little vibration, and it seemed like the building swayed a little. Then it just dropped straight down. It just crumbled right down to the ground. Somebody yelled, 'Run!' and we turned around and ran. It was a while before they let us go back on the bucket brigade. We were there until about midnight.

"We came back to the firehouse, and there were people all over the place. We pretty much had a full house. We had a covering FDNY engine because so many firefighters in our battalion were missing. We lost a lot of guys from Ladder 118, we lost a lot from Ladder 101, we lost Engine 279, and we lost a couple guys from Engine 226. The only two engine companies in our battalion that didn't lose anybody were Engine 204 and Engine 224.

"We also had a truck from New Jersey that was covering the area, and it was parked outside for two nights. Everybody wanted to help. I told the guys, 'There are beds upstairs. Take blankets. We have extra mattresses downstairs, just take them. We'll worry about it in the morning.'

"I stayed at the firehouse that night, and some time around 4 A.M., I finally lay down. I actually took my bunker jacket and put it on the floor behind the covering rig, and that's where I slept. Right behind the rig. At this point, I didn't really care where I lay down. I just put my coat on the floor and slept right there.

"They came down with a list of missing people. I saw a couple of my friends' names on there, and I was still saying to myself, *They're going to find these people. They're just sitting there, they're just hanging out, they're waiting for us to come get them. They may be a little broken up, but we're gonna find them.* I was scrolling down the list, looking at the names of people I recognized, and saying, 'Oh man, why did you have to be working that day?' But then again, somebody had to be working that day.

"I had a friend on Engine 279, Ronnie Henderson. I didn't actually know he was one of the guys who were missing because I guess I skipped over his name on the list. Everyone met at Engine 211 again the next day, and I asked a guy from his company, 'Where's Ronnie?' He said, 'We're going out there to get him.' I said, 'Ronnie can't be gone. Just last week we were playing basketball.'

"The second day, some of the names came off the list. I was just praying, 'Oh please, let Ronnie's name come off the list.' I was praying that Karl's name would come off, and Hector's, and some of the other guys from my battalion. We went out there looking for them every day for a week. The bucket brigade went on and on. People were passing buckets up onto a mountain of debris and passing them back down again. I was looking at it and thinking, *How long is this going to take?* It seemed like we weren't getting anywhere. I'd get so impatient, I just wanted to scream, 'I can't do this anymore! I've gotta do something else!'

"One night when we were down at Ground Zero, we saw our rig. We went into a building, up the stairs, and walked right out the window onto the rubble. Our rig was just sitting there, in the middle of everything. It was broken in half, just bent in half and sitting in the pile. The front part of it was intact, but the entire hose bed was cracked and broken in half. It was covered with dust. We opened the doors and a bunch of dirt fell out.

"It's not normal to walk out of a second-story window and see your rig, which had been parked on the street. There was rubble under it and rubble piled up all around it, and there was the rig, sitting in the middle like it was just waiting for us to come out. I still can't understand it. How could the rig be sitting there with so much debris under it?

"Every day they were coming down with a more accurate list of the guys who were definitely missing. They said, 'These are the guys who came to work, and we cannot find them. These are the members who are unaccounted for.' Even as time went on, I was still thinking they'd find them. I'd think, *Maybe they're in one of those pockets.* Even after two months went by, I kept saying, 'They'll find them. They'll just be sitting there, waiting for us. They'll all be together, and they'll be sitting down in the bottom of this place. They've probably found water, they're probably munching on candy and just hanging out down there.'

"I spent the month of November down at Ground Zero, digging in the dirt. Sometimes I was moving big objects by hand, but otherwise I had a little shovel and was just digging and digging. I had a little shovel and was in the hole where the Towers once stood, and I was just digging and digging.

"While I was there, they found two bodies. I said to myself, *I don't want to come across these guys' bodies because I want to remember them the way I last saw them.*

"We dug up crushed fire extinguishers and mangled air packs. Every time someone would find a void, it brought hope, like maybe somebody's under there. There are times when I actually had to watch the cranes. They'd pick up a load, and I'd look as they moved it to see what was there. That just made me dizzy. I'd be standing there for an hour or two hours, just watching the cranes.

"Even in late November, there were still fires burning underneath the rubble. I'd see smoke coming up, little spots of smoke, and they would tell us to take the hose over and keep spraying water. We'd spray water for 30 or 40 minutes. I'd see the smoke coming up and disappearing, everywhere I looked. When I came back after that month, I had bronchitis. I had conjunctivitis. I had pharyngitis. I've never been that sick.

"When I tell people I was in that hole digging where the Twin Towers used to stand, I still can't believe it myself. I think that one day I'm going to wake up. But everybody can't be having the same dream. I can't wake up and have everything be back to normal. What's normal now, anyway? We can't forget this. It's not like a little fire where somebody got hurt. We lost 343 members of the FDNY. That's a lot of people.

"I keep in touch with the families of a couple of the guys we lost. I've known one guy, Karl Joseph, for 11 years. We took the test together. We went through the academy together. We came out and graduated together. I try to call his mother at least once every two weeks to see if she needs anything—if she needs any groceries, if she just wants to come over and talk—just to see if everything is fine. My parents call, too, because they live nearby. Sometimes my mom cooks food and takes it over to Karl's family. We try to help them out any way we can.

"I've learned not to take anything for granted. Just because people are here one day, they may not be here tomorrow. Two of my friends were arguing with each other, and I said to them, 'After September 11, you guys are going to sit here and say you're not going to speak to each other? If he goes to work and doesn't come home, what are you going to say then?' It really makes you think. After losing so many friends at Ground Zero, nobody has to say that to me.

"The guys I work with at Engine 204 are great. I can't complain. They always look out for me and make sure I have everything I need. Especially when I first came on, everything was new to me. I didn't know what to expect. They always said, 'If you need anything, just call us.' They've made me feel very comfortable here.

"I definitely would encourage other women to become firefighters. When I was young, I wanted to do so many things. I wanted to be a pilot. I wanted to be an auto mechanic. I wanted to be a cop. A lot of the things I wanted to do, my mother said, 'You can't do that. Girls don't do that.' After a while I started thinking, *Well, gee. Everything I want to do, it seems that girls just don't do those things.*

"Hopefully I'll have kids someday, and I'll tell my daughter, 'Don't let anyone ever tell you that you can't do this or you can't do that. You can do it! You can do anything you want to do! If the guys can do it, there's no reason why you can't do it, too. I did it. And look at all the other women who've done it, too.'"

LT MONROE
FDNY
FIRE DEPARTMENT
CITY OF
NEW YORK

LIEUTENANT AMY MONROE

AGE 42
EMS LIEUTENANT
FIRE DEPARTMENT OF NEW YORK
EMERGENCY MEDICAL SERVICES COMMAND
BATTALION 4
LOWER MANHATTAN

Lieutenant Amy Monroe secured a quiet conference room at FDNY headquarters, offered us coffee, and seated herself at the head of the table.

"I'm going to tell you very matter-of-factly what happened," she said, "but I can't even begin to get emotional about it. I've run this story through my mind a lot, and I'm sort of finished with it."

Scarcely were we into the first 10 minutes of our interview, however, before Amy's emotions rose to the surface. It was especially hard for her to talk about her two young sons and the very real possibility that they could have been left motherless on September 11.

Shortly after the terrorist attacks, Amy consented to a lengthy interview with producers of a two-hour network special that aired on TV a month later. Although Amy and two other female paramedics were interviewed for the same amount of time as their male counterparts, their collective comments were reduced to only a few a minutes of airtime.

"The men described what they did," she said, "and the women said how they felt. I don't know if that's why our stories were cut, but I do know that it's very difficult to give this story away to people because it's so personal. What's even more difficult is giving it away and having it be discarded."

"The only radio transmission I heard was, 'Deploy the antidote kits for weapons of mass destruction!'"

—Lieutenant Amy Monroe, October 21, 2001

"The Emergency Medical Services component of the Fire Department of New York is battalion-based, which means we run out of stations. I've been a paramedic since 1991, and I'm a lieutenant at Battalion 4 down on South Street and Clinton. It's the only EMS battalion in Lower Manhattan, and that's our primary area of response. We run about 120 people and 10 or 11 ambulances.

"Approximately 35 percent of EMS personnel are women. I did a lot of my time as a paramedic and as a boss in the South Bronx and Harlem, where there are a lot of women. Here at Battalion 4, for some reason, there are only three or four women. Until recently, there were no female supervisors here at all.

"September 11 was my day off. I had slept late, my husband had just left to take the kids to school, and I had turned on the news and saw that this plane had crashed into the World Trade Center. One of the reasons I went to Battalion 4 was to get experience in supervising Mass Casualty Incidents, or MCIs. A lot of big things happen in that battalion because of the tall buildings in Lower Manhattan, and we run those kinds of jobs all the time. It doesn't mean that thousands of people die, but certainly we do jobs where we treat 30, 50, or 70 patients at once.

"We have everything from gas leak exposures to high-rise fires. We have train incidents, bus accidents, things that happen with the ferryboats out on the water—a lot of different kinds of jobs. Six months earlier, I'd supervised an MCI at Five World Trade that involved a big fire. With any event at the World Trade Center, there are so many people. That's the bottom line. It's just mayhem. Plus, there was a big scare in New York with the World Trade Center bombing in 1993, and people remember that. My partner at Battalion 4 was the first boss arriving at that incident, so I know a lot about the World Trade Center.

"I didn't want to overreact to what I was seeing on TV, because I take pride in not getting too excited about things like this. I know the size of the building, and I was thinking, *It's probably bad, but not that bad.* I picked up the phone and called the station. A boss answered the phone, screaming, 'Just get in here! Come in! Come in!'

"I grabbed my clothes, got in my car, and drove to my kids' school, which is about three blocks away. I ran in to find my husband. I wanted to say, 'Look, this is what's happening, this is where I'm going.' It wasn't that much out of the way to do that. I couldn't find him, so I just jumped back in the car and drove to the station. There were only two people there, both light-duty people who'd been injured, because everybody else had already gone. I asked one of the guys, 'Can you drive?' He said, 'Yeah, I can drive,' so we got into an ambulance, and on the way there I was on my cell phone constantly, calling my husband. I never do this. I don't even really talk about jobs I do unless they're extremely humorous.

"I kept trying to call him, but I couldn't get through. What I was going to say to him was, 'I've got a bad feeling about this.' By then, I knew that the second plane had struck, and I was absolutely sure it was terrorists. I was going to say, 'I don't know if I'm coming back.' I had a really bad gut feeling about it.

"I used to do training for all the firehouses in that area to teach the firefighters about weapons of mass destruction and the antidote kits we carry. We always used the World Trade Center as an example because we have so many jobs there with fumes or strange odors. I'd tell the firefighters, 'You really need to stop and think about what kind of situation you're running into.' I've been in every firehouse downtown. I knew all these firefighters, and this was sort of my thing with them. I'd say, 'You know, you guys gotta stop smelling this stuff in the trash can,' because they'd just go up and smell it and say, 'Hmmm. Smells like this or that,' not knowing what it could be.

"When I got down there, I ran into EMS Captain Janice Olszewski and became the transport officer on Church and Fulton in front of One World Trade, which is the North Tower. We were trying to save the ambulances for people who were really hurt, because that's what you want to do at an MCI. We had a lot of patients who were piling up against the Millennium Hotel, but we were trying to get the burn patients out of there right away. At the same time, we knew we didn't have anybody yet who had been even close to those top floors. They hadn't had enough time to get down. It takes a tremendous amount of time to come down 80 floors, especially when the stairwells are crowded. So we were trying to save the ambulances for them.

"I saw a woman pass out in the street. I reached down to help somebody drag her out of the street, and I heard this really, really, really terrible noise. It sounded to me like an aircraft, and I thought they were sending in another plane to wipe out the rescue workers. I also sit on the Dignitary Protection Unit, which does the security when the president or vice president comes to town. One of the things the FBI and CIA are always warning us about is watching for secondary incendiary devices.

"Then I see this huge, huge crowd running toward us, and Janice said, 'Run!' so I started running. I'm in pretty good shape because I run a lot, and I thought, *I can't believe I can't run any faster than this.* It turned out that I was actually running uphill, so it seemed really slow to me. I said to Janice, 'I'm scared. I'm very, very scared.' And she said, 'Well, run faster.'

"So we reached out to grab hands, and we ran up the street, and this crowd just sort of overtook us and we got separated. I looked down and saw a cop standing in the subway facing me, looking at what was coming, and there was horror on his face. I had my back to it, so I didn't know what was coming. I thought I could either jump under something or I could run down into the subway. But I didn't want to be buried. I didn't want to be trapped. I thought, *If I'm going to die, I don't want to die this way.*

"I was running alongside St. Paul's Chapel, and if you weren't religious before, you would be then, because for some reason, that church was not touched. It's the oldest church in Manhattan, and not one window was broken. The church has a huge wrought-iron fence around it, and I grabbed on to that when the wind hit and everything went totally black. I mean, it wasn't gray, it was black. At that point I really, really, really thought I was going to die. It got absolutely black. I started praying, 'Oh God, don't let me die now, because I have two kids.' That was my chief concern.

"In that moment, I felt sort of stupid for doing this job. It's one thing to do a dangerous job because you really love it, when you don't have kids. But when you have children, things are a little different. My boys are still little. They're only four and six. I had 'buffed' this job, which means that I volunteered for it that day. I wasn't scheduled to be at work. But when you do this kind of work, you do it. You don't think about it, you just go. I kept asking myself, *What in the hell are you doing here? Now you're going to die. You're going to leave the kids alone. What a selfish, selfish thing.*

"Then I saw a light. I kid you not, because this doesn't sound real, but there was this bright orange light. I saw it, even though I had my eyes shut. I thought, *Is this the light they talk about seeing when you die?* I reached down and pulled my shirt over my face and started to breathe through my shirt because the air was so thick it was like somebody was stuffing cotton balls down my throat. I could hear all these voices around me saying, 'Help me. Please help me. I'm dying. I can't breathe.' People were just saying that, out of the dark. They weren't screaming. They were just very soft-spoken. It was really, really eerie.

"I opened my eyes, and things were more gray. I looked around and saw an outline of a vehicle, and it almost didn't look like a car because it had ash on it about three inches thick. I found the door handle, opened it, and got in. There were three people inside. There was a doctor who looked like he was about 22 in the back seat, a macho guy in his early 40s in the driver's seat, and an older man about 70 on the passenger side. He scooted over so I could get in.

"This car turned out to be a Suburban. In the fire department, all the bosses, chiefs, and lieutenants drive Suburbans. This is what I normally drive. It wasn't an EMS vehicle, so I knew it had to belong to a fire officer. I looked over and saw all these wires sticking out from under the dash, and I said to the guy in the driver's seat, 'What are you doing?' He said, 'I'm trying to hotwire it.'

"I sort of composed myself, and then I said, 'Look. I'm Lieutenant Amy Monroe. I work for FDNY-EMS. Go ahead and get this thing going. Know right now that we're going to survive because we're getting out of here.' He just broke down crying.

"It was getting a little lighter, and I looked around and saw another Suburban with its lights on. I asked this guy to go out and see if the doors were open. In the meantime, I realized that I still had my radio with me. The only radio transmission I heard was, 'Deploy the antidote kits for weapons of mass destruction.'

"It's not a secret, but most people don't know that we've been carrying these things for years. Ever since the 1993 World Trade Center bombing, the federal government has mandated that we carry these kits, and there's a huge deployment of them in the city. The kits contain injections you give for anthrax exposure in huge quantities or other kinds of biological exposures. If you see patients who are experiencing overt symptoms secondary to an exposure, you're supposed to inject them with this antidote.

"Then I was thinking, *What the hell is this stuff we have all over us? Was it a bomb that went off on the plane?* I'd figured out by then that the Towers had collapsed, but I had no idea at the time what had made them collapse. I was thinking that it was from a bomb, but what kind of a bomb? Or maybe a nuclear device. In any case, what had we been exposed to? They were screaming over the radio to deploy the kits, and I was thinking, *We have this stuff all over us. We have just minutes to get to a hospital.*

"I was pretty panicked. I felt like I was in control emotionally, but I was thinking, *We've got to get out of here.* The guy I'd sent out to check the doors on the other Suburban came back crying, throwing up, and very emotional. He couldn't breathe, and he was out of control. He told me that the doors were locked. So I said to the doctor in the back seat, 'Is there anything back there, like a tool or anything?' He reached back and grabbed this big drill and said, 'Will this do?' I said, 'Yeah. Get out. Let's go.'

"We broke the window but still couldn't get the door open. We crawled through the window, and I cut my hands all up. The guys got in the back seat, and I started driving to the hospital. I swear to God, we were driving through blackness and suddenly we came out into bright sunshine. People in the street didn't really know what had happened.

"Then we were in Manhattan traffic, trying to get to a hospital and this guy was throwing up. I was thinking, *When are the rest of us going to start having these symptoms?* I am actually the only EMS woman in this city who is on the FEMA Urban Search and Rescue Team, and because I'd had a lot of training in hazardous materials, I knew a lot about this stuff. I didn't know if we'd been exposed to something or not, but this guy was having all these symptoms, and he was really out of control. I kept thinking, *This guy over here is going downhill fast.*

"I could feel the vehicle dying, and I knew how far we still had to go. I'm also an emergency room nurse, and I was going to Beth Israel Hospital, where I worked as a nurse. That happened to be the closest hospital. I passed another EMS supervisor in his Suburban who had some other people with him, and he looked at me. Obviously, he hadn't been in the collapse, because his car was clean. So he started following me. I made a turn about a mile from the hospital, the car died, and he pulled up right behind us.

"He had four patients in his car, and they all had minor injuries. I said, 'Let your patients go to this clinic over here. I have all these people who have been in the collapse.' By then this guy who was with us was rolling around and vomiting in the middle of 14th Street—which is a major street—unable to breathe. The people on the street were bringing us water because we were spitting up all this black stuff. I said, 'We've got to get these guys to the hospital.'

"So the other EMS supervisor put all my patients in his car, and we drove like that, all of us jam-packed in the car, about 10 people, to the hospital. I literally sat in his lap while he drove.

"When we got to the hospital, they put me in the acute treatment area, and they had about 10 doctors on me. I said, 'Wait a minute, do you guys have the antidote kits?' A lot of times the hospitals don't have them. I said, 'Don't BS me, because we've all been exposed.' They looked at me and said, 'Then you guys have to be decontaminated.' Of course, we'd already contaminated the whole hospital, but they sent us back outside where they'd set up a decon station on the street. I went behind a curtain, stripped, deconned myself with water, and went back in.

"I was having trouble breathing, and they sent all these people in to interview me because they were trying to figure out if it was biological warfare. A friend of mine who works there asked me, 'What do you want to do?' I said, 'I just want to get my head together,' because the whole time I was going down there, I thought that Janice was dead. I kept saying her name, 'Janice, Janice.' I was telling the supervisor who was driving us, 'Janice Olszewski's dead. Everybody I was with is dead.' That's what I really believed in my mind. I was starting to have that 'Is this a dream? Is this not a dream?' feeling, and I really needed to get a hold of myself mentally. The charge nurse said, 'Let me give you a little bit of anti-anxiety medication.' I said 'Fine.' A few minutes later she said, 'Well, they just bombed the Pentagon.'

"At this point, I guess I realized that I hadn't been exposed to anything and I wanted go back down to the site and see what I could do to help. I pulled the IV out of my arm and decided to leave. I didn't have any clothes, they had the street blocked off, and everything was out of control. They wouldn't let me off the block because I was wearing a hospital gown and booties. They probably just looked at me and thought, *She's a psychiatric patient.* I walked over to a cop and said, 'Look, I work with the fire department,' and he said, 'Go.' So they let me through.

"I walked five blocks home in my hospital gown, went upstairs, took a shower, got dressed, and went back downstairs. My car was still at the battalion, and all the streets and bridges were shut down, so I went to my neighbor downstairs who has a motorcycle and said, 'Put me on the back of your motorcycle.' They let us on the highway because I had my fire department jacket on, and we took the highway, which was really weird, because it was the middle of the day and everything was shut down. Everything. We went to the battalion, I got in an ambulance with about 10 other people, and we drove back down to the site.

"The first time I'd gone down there that morning, I was really obsessed with my equipment. I had taken—and I have never, ever taken this to a job—all my Urban Search and Rescue gear, my mask for the airpack and everything, but we'd had to leave it in the ambulance. I told the guy who drove me there, 'You've got to hide this.' I thought I might need to use my gear for a USAR deployment, and I figured if things got crazy, everything was going to disappear. I never do this either, but I made him leave the keys in the ignition.

"So when I went down there the second time, I had somebody take me in an ambulance to try and find this vehicle. It had been abandoned, and the keys were still in the ignition. I went into the back, and everything had been stripped out of the ambulance. But he had hidden my USAR gear in a compartment, and I found it. I took it and went down to deploy with the FEMA team.

"At that point, I had ceased to be working in my usual capacity as an EMS supervisor and began working on the Urban Search and Rescue team for the federal government.

"Everything was pretty out of control that night. It was insanity. Working for the fire department, you have such safety concerns. But that night, absolutely not. Bulldozers were just flying by, cranes were coming in, we were walking over huge things, and there were live wires everywhere. It was bizarre. We'd be sitting on a windowsill of a building and look up, and there'd be huge shards of glass hanging over our heads. Everything was blown out.

"There were firefighters out there, desperately digging into the pockets, and there were volunteers out there who were about 18 years old, in their jeans and sneakers, trying to do the right thing. People were digging with their bare hands. It was the most terrible thing that had ever happened, and we were desperate to find people.

"The bodies they found were just like mush. A lot of body parts. I was drinking a bottle of water, and everything was really crazy, really really crazy, and I looked down at my foot and saw this woman's hand with a wedding ring on it. Then one of the volunteers came and scooped it up. I've seen jumpers and I've seen children beaten to death and stuff like that, but this was really, really different. I knew that where that one hand was, there were thousands and thousands more. I didn't know anything about this person, except for the color of her skin and the fact that she was married. Obviously, she had a connection, she was loved, she was part of a family. This one hand represented many, many people who would be hurt because this one person had died.

"It was so strange there that night, because the lights were on, and the site, even to this day, looks a lot different at night. Nighttime is sort of a peaceful time in a way, and they used these really bright searchlights to light the place, so it created a surrealistic atmosphere. The devastation is hard to comprehend even when you look at it in the daytime. But at night, it's very different. It wasn't like there were ghosts there, but it was still sort of alive.

"My assignment was to provide medical care to people who they found trapped, hook them up to an IV, and give medical support. We were waiting to be told that someone had been found, and it was very frustrating because they weren't finding anybody. There was nobody to help. That can be pretty frustrating when that's what you do for a living. That probably contributes as much to post-traumatic stress as anything.

"I was down there for three days with FEMA, then I went to work on the Operation side. Then I didn't go back for a really long time. I had situations that forced me to be in that area. I had to walk by, and, of course, I couldn't walk by without looking. There's still a really weird energy down there. Maybe it's because there are so many souls.

"I try not to be overly paranoid, and I'm a very optimistic person, but I have to tell you, next time we're going to lose our EMS people if it's a biological attack. It won't come in as a fire. It will come in as a medical call, and our EMS people will be the first ones on scene. It's not going to be the white powder in the envelope. It's going to be the big thing that happens, and it will wipe out a lot of people. Do I believe that can happen? I will never, ever, ever think that anything is impossible after seeing what happened on September 11. I will never think that anything is beyond the realm of comprehension after that.

"Everybody asks how my children reacted. My four-year-old really doesn't understand. At first my six-year-old was asking, 'Why are people fighting?' Then it was, 'I'm never going to get on an airplane because I'm afraid it's going to crash.'

"One of the things I get emotional about is that there have been literally, from across this country, thousands and thousands and thousands of letters sent from children to firefighters, EMS personnel, and police officers in New York City. I've got a box of 400 letters sitting behind my desk right now. I can't bear not to respond, but we can't respond to every kid. These letters are displayed all over the city, in firehouses, police departments—everywhere.

"This has changed my children's lives forever. They will never grow up with the security I had growing up. The world has been changed, and that pisses me off. That really pisses me off, more than anything. But on the other hand, I look at these letters from children, and it touches me that they'd reach out to us like that. My kids go to school downtown, and children from across the country have sent them stuffed animals. They come home with these things and I think, *Oh my God, my kids are considered victims.* And they really are. All children in this country are victims.

"We lost eight paramedics and EMTs on September 11. What the people in EMS did that day was to go down there to help people, knowing that it was dangerous, knowing that we could die. Janice volunteered to go, I volunteered to go, and a lot of other women volunteered to go. There is a perception that women don't do dangerous things. The truth is that women *do* do dangerous things. We just do them differently than men do sometimes.

"I think it's really important for the little girls in this country to read this book and see that women are not afraid to do dangerous things. I think this book will be a great example for children, because women are still very stigmatized. This is still perceived as being a man's job.

"I live in a small condominium complex with 12 units, and we all know each other. We all have children, and there are 14 kids who live there. When I came home from work after being out on the pile all night, the kids had this sign up that said, 'You are my hero.' The little girls wrote that to me. One girl wrote, 'Thank you for letting me know that women can do anything.'

"I want to tell those little girls, 'You can do anything, too. You can be anything you want to be.'"

POLICE
NEW YORK
ROYSTER

LIEUTENANT KIM ROYSTER

AGE 39
LIEUTENANT
NEW YORK CITY POLICE DEPARTMENT
DEPUTY COMMISSIONER'S OFFICE OF
PUBLIC INFORMATION

Kim Royster was a sergeant with the NYPD when we interviewed her in November 2001. She showed us into the conference room at the Deputy Commissioner's Office of Public Information and began to talk. Before we were even settled into our seats, she was talking about Moira Smith, the only female police officer in the NYPD to die on September 11.

"The women of the NYPD are not only heroic and courageous," she went on to say, "but they have such depth. They're intelligent, they're strong, they're powerful, and they're mighty. Yet they're pretty, they're feminine, and they're soft. They have very big hearts. They have hearts of gold."

Kim impressed us with the many hats she wears. She's a mother, wife, police officer, and part of a skilled team of media professionals who handle all press inquiries for the NYPD. She's also an accomplished vocalist who uses her music to express her patriotism, her emotions, and her desire to help others heal their breaking hearts.

After she finished telling us about her own brush with death on September 11, Kim treated us to a few rifts of song. Mindless of the tape recorder reeling away on the table or her colleagues milling about the halls, she broke out with, "Oh say, can you see by the dawn's early light ..." and captivated us with her powerful, moving voice.

Eight months after the attack on the World Trade Center, Kim was promoted to lieutenant and assigned to the 13th Precinct to serve as platoon commander.

"I'm a wife, a mother, a woman, and a police officer. Those four things I will always be."

—Lieutenant Kim Royster, November 2, 2001

"I came to the police department as a civilian employee in 1984. I was working at the Police Academy, where all the new recruits come in for their training, and I decided that one day I was going to become a police officer, too.

"Since then, I've put in 15 years as a police officer with the NYPD. I was first assigned to the 13th Precinct, where I did patrol in the Gramercy Park area. Then I was transferred to the Criminal Justice Bureau. I told myself that once I was a police officer, I was going to rise through the ranks. I made sergeant in 1997, and I should make lieutenant within the next six months. Then I'm going to take the captain's exam.

"In addition to being a police officer, I've always been a vocalist. I've sung at the Apollo Theater, and I do a lot of jazz, rhythm and blues, classical, and Broadway show tunes. I started singing the National Anthem at NYPD promotion ceremonies and inaugurations, and I would sing any time a new police commissioner was sworn in. Any ceremony that was patriotic, I was there. We had a female Deputy Commissioner here at the Public Information Office at that time, and she said, 'Kim, I'd like you to come work in my office and start MCing the ceremonies.' I said, 'I'm a singer. I don't want to talk.' She said, 'Well, just come and give it a try.'

"The Deputy Commissioner's Office of Public Information is a 24-hour-a-day, 7-day-a-week operation. We are citywide, meaning that we handle Brooklyn, the Bronx, Queens, Manhattan, and Staten Island—all five Boroughs. Any police-related information the press wants to know is disseminated from this office. We handed out 4,000 temporary press credentials when the World Trade Center disaster happened. That was in addition to the 10,000 we already have in the system. We have an average of 1,000 press calls coming in every day. We have approximately 35 people working here. Do the math. Have we been busy!

"September 11 was Election Day. All eyes were focused on who was going to be running in the Primaries for the mayoral race and different Borough presidents. I woke up that morning, went into the kitchen, and my husband, Gregory, said, 'Kim, something happened. A plane hit the World Trade Center.' I said, 'Oh, my God!' Right away my heart started beating rapidly, because in 1993 I was in the Public Information Office when the World Trade Center was bombed.

“We jumped into the car and came here like lightening. It was one of the most beautiful days that you would ever see. It was crisp, clear, and warm, and there wasn’t a cloud in the sky. As we were approaching the city on the Brooklyn Bridge, the sky filled with smoke. By then, both Towers were burning. My husband and I looked at each other and the tears started to roll.

“I got to my office, but everyone had been evacuated from the building and it was locked. I found two of our female detectives downstairs, and they told me that everybody had gone to the scene. I said, ‘It’s going to be impossible to drive over there. Come on, we’re walking.’ We got on all our gear and made sure we had everything. Mind you, that day I had on a dress with sandals. We were only about eight blocks away from the World Trade Center, and as we got closer, we started seeing shoes, bags, glass, blood, and people running, screaming, and crying. We knew it was bad.

“We got to the corner of Church and Vesey streets, and there was no longer any way to contact anyone. Our cell phones weren’t working. Our beepers weren’t working. I looked around the corner, and there was a wheel from the airplane. A wheel had fallen right at the corner. I left the detectives there and went to look for the rest of our team.

“I met up with Lieutenant Terri Tobin and told her that we were setting up on the corner of Church and Vesey. We were going to gather the press together there to give them information that needed to get out to the public. We wanted to keep them all together and make sure they were out of the frozen zone so that they wouldn’t get hurt. Terri went to find the rest of our team, and that was the last I saw of her that day.

“About two seconds after that, I heard this roaring sound. I looked up, and it seemed like we were being overcome by a big monster. The North Tower was collapsing. People were running, and I started to run, too. All I felt was fear. I was choked up with it. I thought I was going to die, and I was willing to accept death. I prayed, *If I’m going to go, Lord, please don’t let it hurt.* That’s what I said. I know that God was looking down on me, and it just wasn’t my time. But then later I questioned, ‘Why was it theirs?’ All these other people were sacrificed.

“A gush of smoke engulfed me, and I pushed my way into a building. After I got inside, I realized it was a church. Other people ran in right after me, and I started calling more people in. ‘Come in! Come in! Come into this building! It’s safe in here!’ I guess I got about 40 people into the building. We looked outside, and it was pitch black.

“I’ve never been to hell. I never want to go to hell. But that’s what it reminded me of. Hell. Dark. Stifling. You couldn’t breathe. You couldn’t see. You didn’t know where you were.

"People were talking and crying because they were afraid, and I said to everyone, 'Be quiet now. We have to think smart.' I asked the church security guard, 'Do you have a basement?' He showed me where it was, and I said, 'Come on, everybody. Let's go down to the basement.' We got down there, and I said, 'Let's just gather ourselves. We have to think about what to do next.'

"Two Emergency Services guys came in, and I couldn't see anything but their eyes. They were covered with white ash, and their eyes were red. I asked the security guard for some water, and I helped them wash their eyes out. Then they went right back out.

"I thought, *We need an exit going north. We cannot go south.* We finally found an exit going north, and I got everybody out. I said, 'Come on, go! Just keep going, just keep moving.' I said to myself, *I can't go back, because I don't know where anyone is.* I decided to go to the office to try and coordinate things from there. It was a very long day.

"That night, I was driving home over the Brooklyn Bridge, and it was quiet. Quiet like a snowstorm. It was the most eerie feeling I've ever had. Usually, that bridge is packed with people, but there was no one on the bridge. I had the radio on, and I heard Whitney Houston singing the National Anthem. I heard that song 'Proud to Be an American,' and I heard 'America the Beautiful.' I cried all the way home.

"I am in no way a hero. I had the experience of being overcome by the smoke, but I can't imagine what my brothers and sisters dealt with in that inferno. I didn't rescue anybody. I didn't evacuate anyone. I was just there to experience it, and I never want to experience anything like that again in my life. Those people who died—the 343 firefighters, the 23 NYPD officers, the 37 Port Authority officers, and the 6 people from voluntary EMS agencies—they are the heroes. There isn't a word that can explain or tell you how heroic they were. They lost their lives, but they also evacuated thousands of people from those buildings before they collapsed.

"Throughout this whole tragedy, I've had the opportunity to interview many people who were involved in the evacuation and the rescue and recovery efforts. I've had the opportunity to listen to their stories. What brave men and women we have! There were people who went into those staircases and felt the first Tower go down but would not leave until everybody was out.

"Cops and firefighters have that instinct: *This is what I've got to do, I'm going to do it, and I'm going to make sure I do it well.* You don't think to yourself that you could get hurt. Those people didn't think that. They put that out of their minds. *I've got to get these people out of there. They need me.* That's what they were thinking.

"People came from everywhere to help. We had the biggest gang you would ever want to see. I can't thank people enough. They came from all over the United States to help find our brothers and sisters. They were out there 24 hours a day, 7 days a week. They wouldn't stop. In fact, Rudy Giuliani, who was our mayor at the time, had to make them stop.

"This changed our lives. We came back, and we worked and worked and worked and worked to forget about what happened. That's what we were doing—trying to forget about it. We wanted to work. What else could we do? Calls to our office tripled. Press from all over the world was coming here. They wanted to interview the officers. They wanted to put them on their TV shows. People wanted to donate money; they wanted to go down and help. We had to make sure we regulated what was going on. We were everywhere.

"The day after the tragedy, we went out to Ground Zero because everyone from the media wanted to get pictures and tell the story. People were down there digging through the rubble, troops were coming in, and volunteers were there to help. We saw emergency vehicles totally destroyed—flattened and blown to pieces. We wanted to let the world know: 'Look what happened to us. Look what these heartless people did to us. Look what hatred did—tried to ruin New York City; tried to ruin America.'

"I didn't cry. That scared me. I didn't cry, and we just kept working. On Saturday night I went home and my son, Joseph, asked me, 'Are you okay? You've got this look on your face.' I said, 'I'm fine.' Then he said something about the flowers. I have a garden, and I like my garden, and I just broke down and started crying.

"My husband embraced me and said, 'Let it out. You have to.' And I said, 'Thousands of people died! I don't understand why!' And I just cried and cried and cried.

"I wasn't a hero, but this opened me up as far as being American, loving my country, and having pride for myself. I think it made me a better person. I work harder to make things better for people, to make people laugh, to make people feel comfortable. I don't want to be a person who could be looked at as negative, as having negative energy. Life is short, I realize. Your life can be snuffed out in a second.

"This has made me appreciate my freedom a whole lot more. I wake up every day and I thank God that I'm alive and that my family's okay. I think about life and living it to the fullest. I let all the trivial things go by the wayside. I'm nicer. I say, 'Please,' and 'Thank you,' and 'I'll hold the door.' And people are nicer to me, I've noticed, because we all realize what happened. I can't walk by a firehouse without crying. I don't rush anymore. Sometimes I used to rush—just rush, rush, rush. It's okay. It'll work itself out.

"And baseball games. If people weren't baseball fans before, they are now, just because it's as American as apple pie. 'Yeah Yankees!' It means New York. It's good that we won the American League Pennant. We needed that. We needed to win. The New York Yankees have won the Pennant before, but it meant nothing until we won it this time. It's American—baseball, that eagle flying across the sky, the flag. I have a flag out at my house—a very large one—and I won't take it down until our troops come home.

"My husband told me, 'If I could, I'd take you away from this. My wife should not be involved in anything like this.' I said, 'But you have forgotten one thing. I'm a wife, a mother, a woman, and a police officer. Those four things I will always be.'

"And those titles are very large. I'll always be a police officer, even when I retire. When I was growing up, I used to say that all I wanted was to be the best woman I could be and the best wife and mother I could be. That was very important to me. Now I also want to be the best police officer I can be.

"There's not a day that goes by when I don't think about this. It's really affected a lot of things in New York. The subways were out. Businesses and restaurants were closed, and people weren't going out. Shopping, traffic, the stock market—a lot of things were affected. It's affected me, too.

"Now, when I walk into a room, I make sure I know where the exits are. If I'm on the train and it stops for too long, I get a little impatient. I'm more aware of what's going on around me when I'm driving or walking. I make sure I have my weapon with me all the time. I think twice about conversations I have with people. What are they talking about? What do they really think?

"But we can't walk around with fear. God's blessed me with a wonderful son. He's 22 years old, he just finished college in May, and he's doing an internship with the Red Cross here in New York. September 11 tore him to pieces. I told him, 'New Yorkers have this way of not fearing anything. We're rock hard, but we're definitely soft inside, too. We are proud and strong, and yet we were hurt. We have to go on with our lives. Our fallen brothers and sisters would want that. So don't back down. Don't give in to the fear. Don't give them the satisfaction of knowing you're afraid.'

"The Sunday night after this happened, I was asked to sing the National Anthem at the prayer vigil at Yankee Stadium. Thousands of people were expected to be there. I went to that prayer vigil on a mission. My mission was to let everything I had inside me come out through my music. I felt as light as a feather. I was not in control of my voice, which was a wonderful thing, because I was singing from my heart, not from my head.

"That night, I have never felt more like an American. I was an American from the day I was born, but I was truly an American right then and there, and I felt a sense of unity with everyone who was in that stadium. We were all brothers and sisters. We were all Americans. That was the first time I've ever felt that there was no separation, no disparity, no discrimination, no man or woman, and no black or white. We were all Americans. That's what I felt, which was a good feeling. Wouldn't that be a wonderful feeling to have for the rest of your life?

"When I finished singing that song, I had closure. I felt that I'd done what I needed to do to help everybody get through this. All the thousands of people who were out there felt so proud. Families were broken and torn down, and some of the families cried through the whole vigil. All those people who died, all the civilians, the police officers, the firefighters, the rescue workers—they all have wings now. They all have wings."

POLICE DEPARTMENT
NEW YORK
4603

PATTY LUCCI

AGE 27
POLICE OFFICER
NEW YORK CITY POLICE DEPARTMENT
113TH PRECINCT
JAMAICA, QUEENS

Patty Lucci told us over and over again that she didn't think she had much of a story. "I don't think I'm a hero," she said, "and I definitely don't think I've done enough."

But the pictures she painted with her words and the images she shared with us as we sat at the kitchen table in her modest Long Island apartment were among the most vivid portrayals of September 11 that we heard.

Growing up in Syosset, New York, Patty always wanted to be a police officer. "I've always wanted to help people," she said. "I think it's just part of who I am."

For someone so young and barely three years into a career she loves, Patty seems to have the right combination of qualities that make for a good police officer. While her demeanor is one of kindness, compassion, and resilience, her story reveals a stoic, capable woman who showed incredible determination and strength under circumstances that most of us could never imagine.

Even after we'd talked for nearly two hours, Patty still worried that she hadn't been able to properly explain her feelings. "I think you can see what I want to say more than I can express it," she said. "I think it's in my face. I wear it on my sleeve."

As we packed up to leave Patty's apartment, she made one last humble remark. "My story is very simple," she said. "It's just a story of wanting."

"I wanted a chance to find that one thing. A wedding ring. A wallet. If I could have found that one thing that would have given somebody closure, it would have been enough for the rest of my life. That's why I had to dig."

—Patty Lucci, November 3, 2001

"I was sleeping, and I got a phone call from a friend of mine who said to turn on the TV. The first Tower was on fire already, and then I saw the second plane hit. The first thing I thought was, *Oh my God! All those people!* It was devastating.

"Obviously, I knew I'd be going in to work that day, but in what capacity, I didn't know. I got a phone call about an hour later to say that I'd been mobilized. All I felt was fear for the safety of everybody who was going down there.

"The first night, they needed us for the outer perimeter. They put us on a bus, and as we were driving through the city toward Ground Zero, the officers directing traffic were covered in dust and soot. The ground was covered. The city was covered. It looked like a war zone. Even before that, even coming over the bridge into the city, just seeing the smoke coming up and the Towers not being there, you could have heard a pin drop on that bus. All I heard were people's heartbeats. That's all I heard.

"I happened to be stationed over by Ten House, the firehouse across from the World Trade Center that lost the most firefighters. One of the worst things for me was seeing all those men cry. There wasn't anybody who walked into that house who was not crying. They were trying to be so good to us as officers, and we were trying to be so good to them as firefighters, because their loss was incredible. But it got to a point where I couldn't take it. I couldn't go in there anymore.

"Watching the National Guard come into New York, just watching them move through my hometown and knowing why they were there, was amazing. The State Troopers were rushing in, and all the firefighters were coming in from Long Island, Connecticut, New Jersey, and the surrounding areas. I just kind of stepped back and thought, *This is it. This is for real.*

"That morning, after we set up the perimeter and shut down the city, a woman came up to me. She was holding a picture of a friend of hers who was supposedly in the Towers. She was kind of walking around in a daze, and I said, 'Are you okay?' She showed me the picture and said, 'Have you seen her? Have you seen her?'

"And my heart broke. I hadn't seen her friend. I spoke to her for a few minutes and tried to reassure her. I saw the loss in her eyes, the despair, the not knowing. She told me that she'd been walking around all night, that she hadn't been to sleep yet, that she hadn't eaten, and that she couldn't do these things until she found her friend. That was probably one of the most helpless moments I've ever felt in my life, to not be able to give her something. All I could do was hug her. That was it.

"Seeing all those people who just had to come down, seeing them standing behind the barricades, was overwhelming. They all said the same thing: 'I just needed to see it.' I understood that. They would come down at 1 o'clock in the morning, 3 o'clock in the morning, 5 o'clock in the morning, because they just had to be there. A lot of people would cry. A lot of people would just stare in amazement. I saw a lot of empty faces.

"The second night, I got on a bus out of the Precinct with a sergeant and two other cops, and we drove into Manhattan. We loaded up the bus with everything the community had put together for the rescuers: socks, food, underwear, dog food. You name it, they brought it in. Our job was to take it to Ground Zero and distribute it. On the way down, the bus doors were open, and we threw food, socks, and water to any cop or firefighter we saw.

"When we got to the site, we loaded up our bodies with as much as we could. I remember walking up to firefighters and seeing how happy they were to have socks. Just clean socks. They had enough food, but clean socks and underwear made them so happy. It was so simple. It was the most incredible feeling to see how everybody was pulling together, how everybody was just there to help. There was not one selfish soul in the city of New York that night. Not one.

"I looked up and out at what had been the Twin Towers, and I remember feeling numb. Completely helpless. The ground there was buried in dust and paperwork and family photographs and everything you can imagine—personal artifacts, shoes, clothing. All over Manhattan. There were thousands of rescue workers all over the place, trying to do the best they could just to find somebody. The whole block was lined with rescue workers, and they were so tired. A lot of them were crying. I remember thinking, *I am not going to just sit back and do traffic.* I had to be a part of it. I knew I was going to get in there and dig. I had to go in there for my own peace of mind, and there was going to be no stopping me.

"The third night, I was on the other side of the Towers patrolling the perimeter, and every time the wind blew, I smelled the dead bodies. It was an overwhelming scent. Just terrible. It was like we were sitting in a morgue. I felt so much sadness, knowing that it could have been my mother, my father, or my brother. I knew that what I was smelling was somebody who had had life the day before.

"About a block away from Ground Zero, there was an alleyway covered with paperwork from the Twin Towers. It was a dark alleyway with trees, and everything had blown into it. I started looking through it. I didn't know what I was looking for, but I just knew I wanted to find something. I was back there by myself, and it was dark and it was cold. I looked down and there was a photograph. It was partly burned, but I could see that it was a photograph of a family. On the back it said, 'Daddy, I love you.' That's the night I broke down. Right there, by myself, in the alleyway, with all this stuff around me. That was the first night I cried.

"Shutting the city down and securing the perimeter, that's my job. I'll do it any day I'm asked. But I wanted to be in there. I wanted a chance to find that one thing. A wedding ring. A wallet. If I could have found that one thing that would have given somebody closure, it would have been enough for the rest of my life. That's why I had to dig.

"So finally, on my day off, I went into Manhattan and got on one of the bucket lines. There were four lines, with maybe 70 to 100 people on a line. The hole was bigger than my apartment, and it got deeper toward the center. It was just speculation at that point that we were digging for a fire truck. People lined up and passed buckets up the line, and people in the hole filled them and sent them back down again. Person to person, we were just passing buckets. And as I was passing the buckets, what I thought I was seeing was dirt, but it was really dead bodies going by me. I just had to shut myself off to that.

"The longer I was on the line, the closer I got to the hole, because people got tired from digging. I noticed that all the guys were looking at me, and I didn't really know why. Then I saw that I was one of only two females on the bucket line. There was one guy, a little bit smaller than me, passing buckets next to me. When he saw that I was a female, all of a sudden he started saying, 'It's heavy, it's heavy,' every time he handed me a bucket. I took the bucket with one hand and passed it to the next guy and said, 'Okay, now let me know when it's *really* heavy.' He just started laughing, and that was the end of that.

"We were looking for bodies. We were looking for anything. Every shovel full we put into a bucket, we sifted through first with our hands. At some point, they stopped us from digging and kicked us off the line. They'd found the body of a firefighter on the opposite side of the hole, and they were allowing the fire department to come in and recover it.

"I remember walking off thinking that they'd just found somebody's brother. Somebody's son. Somebody's father. The faces of all the guys I was digging with were just empty. Just blank stares. Don't get me wrong, we were happy. I feel so bad using that word, but we were happy to be able to give that firefighter back to his family. But I know we were all thinking the same thing: *Why can't we just find somebody alive?*

"After about two hours on the line, I got up to the hole. The guy below had gotten very fatigued and overheated, and he dropped the shovel. So I jumped in and just started digging. I was throwing buckets, I was calling for buckets, I was just digging my heart out. I dug for a little over an hour. I wasn't sure how the guys would react, but they were great. It wasn't about being male or female. It wasn't about being a cop or a firefighter. It wasn't about being anything but there to help.

"A lot of people put their selfishness and their pride and any and every impure thought or feeling aside. Anything unfair about them was gone. We were all equal. We were all just there for each other. It was like a silent understanding. If you needed something, whether you knew the guy next to you or not, he was going to get it for you. He was going to do it for you. He was going to try and help you through it.

"It was impossible to dig with a respirator on. It was extremely hot, and none of us had our masks on. I was inhaling all that stuff because they were passing buckets over my head and all this stuff was falling on me and the dust was being kicked up. One guy looked down at me and said, 'You should put your mask on.' But the mask just didn't matter to me at that point. That may sound very foolish, but it wasn't my health that mattered. It was getting the people out that mattered the most.

"It got to a point where I became overheated and had to come out of the hole. I was sweating and tired and just covered in soot and dirt and dead bodies and everything. I took a step back and looked up at a neighboring building, and I could see where people had hit it. I could see where they slid down. I remember hearing stories about how many body parts they found on the roof of that building. It was such a weird moment. It was a moment of feeling that I had finally gotten to do something meaningful. It was a moment of just feeling helpless. That was a sight I'll never forget for the rest of my life.

"I walked away from the hole thinking I'd made a difference. But for the amount of time I was down there, for the amount of heart I put into it, when I turned around and looked back, it just didn't look like anything had changed. Those rescue workers down there, the people doing the rescue, recovery, and clean-up, were absolutely amazing. Anybody who doesn't know that it was cleaned up bucket by bucket needs to know the amount of sweat and heart and hard work that went into it.

"I felt guilty showering after I got out of the hole. I wasn't the only one who felt that way, either. A lot of us did. We felt guilty eating. I have a friend who went digging at Ground Zero who couldn't wash her boots afterward. She felt there were people's souls on her boots, and she couldn't bear to wash them off.

"If there's one thing I've learned, it's that tomorrow's not promised. You truly can't take today for granted. This has affected all my friends and family, because with my being a police officer, they all worry about me. There's no reassurance these days. There's none. I think I've brought the people close to me even closer. I tell them I love them more. I try to live every day to it's fullest. And that means living life without regret.

"I would like to think that anybody would have done what we did. We had the opportunity, being police officers, firefighters, and rescue workers, to be able to help. The people of this city and the people who came from all over the country gave us strength. To all the people who said, 'Thank you,' to us, I just want to say, 'Thank you,' to them. It kept our hearts going when we were feeling nothing.

"I hope I never forget the way this city came together. I hope I never forget the way the people of New York were driving around in their personal cars, handing out hot meals they'd cooked on their own stoves for police officers and firefighters and rescue workers. I hope I never forget the way people were coming up to us, asking if we were okay, when we were supposed to be out there looking after them.

"There's more goodness in this world than anything else. Sometimes being a police officer, I forget that, because I'm not always dealing with the best in people. If there's ever a time when I feel I'm going to lose compassion or patience or understanding, that's what I'll remember. The goodness in people."

LOIS MUNGAY

AGE 43
FIREFIGHTER
FIRE DEPARTMENT OF NEW YORK
ENGINE 235
BEDFORD-STUYVESANT, BROOKLYN

Lois Mungay appeared to us as a scrappy, hard-living, hard-working New Yorker with an accent so thick it was clear she'd lived all her life in Queens. This would be a difficult interview, *we told ourselves. As she settled herself into a chair, Lois gave every impression that she would not easily give herself away.*

But beneath that tough exterior, we found a surprising tenderness. In fact, hers turned out to be one of the most emotional interviews we conducted during our tour of New York City.

Lois spoke with great candor, and it was difficult for us not to sob right along with her during this interview. "At least at the firehouse I can talk about this without crying," she said before we were halfway through. "With the guys, I'm okay. But see, women do this to me. They make me break down."

The most decorated female firefighter in the FDNY, Lois may also be the most humble. Someone else had to tell us that she's earned five FDNY unit citations for exceptional performance during her 20-year career.

It hasn't been easy for Lois, being in the first group of 40 female firefighters to join the FDNY after a class-action suit opened the field to women in 1982. But despite the struggle to prove herself, it was clear that she'd come to love her brothers in the fire service. She grieves for them today as if they were members of her own family.

"It was still smoking and the dust was bad and nobody even knew what to do. There were no buckets for a bucket brigade, so we just climbed to the top of the pile and started handing stuff down. That's all we did for 13 hours—handed stuff down."

—Lois Mungay, November 28, 2001

"My alarm clock went off, and I was listening to the radio. They said a plane had crashed into the World Trade Center. I said to myself, *That sounds like the job of a lifetime.* I figured we'd be going on the fifth alarm, so I called the firehouse to say I was coming in early.

"I started toward the firehouse, and then I remembered that the day before I'd picked up pork chops for dinner at the firehouse that night. I thought, *We're not going to get home until the middle of the night. Let me go back and get those pork chops.* So I went back home, which took 10 minutes.

"By the time I got into the firehouse at 9:20 A.M., the rig was gone. I looked at the computer and saw that they'd gone out at 9:10. The rest of us watched it on TV, waiting for the buses to come take us into the city. When the van came, the chief said to me, 'I need you to stay here.' I said, 'What for?' He said, 'I need you in Brooklyn.'

"So I stayed there, and during that time, we watched TV and saw the buildings collapse. A lot of things were going through my mind. When the Towers came down, I thought, *Do I really want to go there now?* I've never walked away from a big job. This was the first time in my career that I thought, *Maybe I don't want to go down there.*

"Then the chauffeur called up the firehouse and said, 'The buildings fell down.' I said, 'Where's the company?' He said, 'I don't know.' I said, 'What're you doing?' He said, 'I'm going to clean the fire truck.' I asked him, 'What the hell are you gonna clean the fire truck for?' He said, 'I don't know.'

"It turned out that he'd had to go two blocks away to get a hydrant because they were having water pressure problems. During that time, the building collapsed. I asked him again, 'Where are the guys?' And he said, 'They're in there.'

"I had seen those buildings come down, and I knew our guys were dead. That was it. I made myself believe that. I wanted to get it over with right then. I thought, *Let me deal with this now. That's the way it is.* No one could have survived that. If they were in that building, they were under there. And when the chauffeur told me where he'd dropped them off and where they went, that was it. He never saw them after that, and he was the only one who came back.

"Our guys were probably in the stairwell, on their way up to the tenth floor of Tower Two when the buildings came down. I don't think anybody knew what the hell to do when they got there. I don't know how you're going to put out 10 floors of fire being fed by jet fuel. I think their only purpose was to evacuate the people and get the hell out. I don't think there were going to be any extinguishments. I doubt that there would have been any rescues. Once they got to where those planes crashed, there was no way they were going to get to those upper floors above the fire. I would assume that the only thing they were told to do—and nobody who was there is around to say—was just evacuate the building and help people down the stairs. That would have been it.

"Later that night we went down there. I was walking with the guys who were held back in Brooklyn with me, and we were about four blocks away when I looked over at this car and saw this big chunk of meat on it. I was an anatomy student, but I couldn't make out this body part. I said, 'Is that a human being?' Someone said, 'I think it's part of a torso.' I thought, *If that's four blocks away, what the hell is down in the hole?*

"So we kept walking, past the Marriott and the other hotel, and they were all beat up. We got to the middle of it, and I looked up and saw it. Fifty feet of rubble. I thought, *Where the hell do we start?*

"I kept asking people, 'Does anybody know where Engine 235 was?' We wanted to start digging where we thought our guys were. I mean, they're all our guys. But if we could have a place to start, we were thinking we might find our guys from Brooklyn.

"It was still smoking and the dust was bad and nobody even knew what to do. We all just looked at each other. There were no buckets for a bucket brigade, so we just climbed to the top of the pile and started handing stuff down. That's all we did for 13 hours—handed stuff down. We didn't even know what building we were on. Then somebody said, 'You're not even on a building yet. You're in the middle of the street.'

"So the only thing we could do, even if it was a little tiny piece of metal, was to pick it up, move it, take a look around, pick up more stuff, and move it. That's what we've been doing for six weeks—picking up stuff and moving it. For the first couple days it was rescues and attempted rescues, but now we're just trying to recover pieces of bodies.

“A couple days after the collapse, we found the body of a lieutenant. We put him in a body bag. He had an ID on him, so we ID'd him. Then we found the body of a civilian next to him, and we scooped it up with shovels and put it in a body bag. After that, we were just finding pieces—a piece of a bunker coat, a piece of a helmet. Any kind of flesh we could find for DNA, we put it in a bag. We didn't know who we were looking at or who we were picking up or what company he was from unless he had a helmet or an ID on him. We just looked for helmet numbers or names on bunker gear, stuff like that.

“Other than that, all we do is stand around. The construction workers move out the big stuff, and then we go check out the voids. Climb down into the hole, see if we can see anything, try to find body parts. If the cranes are moving the heavy stuff, we just watch them dump it into the truck and make sure no body parts fall out. Every now and then, we might see a reflector or something like that standing out in the garbage. Once we see that, we find out where it came from and go over to the area and start searching around.

“Yesterday, they found four guys from a company that was supposed to have been with Engine 235. If they find our guys next, that's where we're going. That's where we'll be. They'll call the firehouse, then we'll start calling each other, and wherever we are, we'll get down there. It's a tradition that firefighters take the bodies and bring them to the morgue. Thirty guys fighting to hold on to a body bag. That's what it's come down to.

“There are funerals or memorials almost every day throughout New York City, but I can't go every day. I just try and go to the ones for the guys I know. I've been on the job almost 20 years, so I know a lot of guys. Not just saying hello, but I personally know more than 60 guys when I go down the list of those who died.

“So far, of the six guys we lost, three of them have had either their funeral or their memorial service. We do everything for the families. We split the company up, and we're all assigned to families. They have questions about pensions, they have questions about insurance, they have questions about taking care of the kids. Money's coming in left and right, and we have to let them know what's going on with that. We escort them to memorials, we bring them down to Victims' Services at the pier, we take food to their houses. We do anything and everything we can for them.

“One guy had three kids, one guy had two kids, one guy had an older kid, one guy didn't have any kids, and the chief had kids and grandchildren. A lot of them want to come down to the firehouse and hang out and talk and eat. That's tough. That's tough to deal with. And they call all the time. ‘I heard this, I heard that.' Then we've got to go find out the truth about what they heard. People are giving them bad advice, and they're asking us. We're not lawyers, so we go find out for them.

"But the worst thing, in the very beginning, was escorting the families to get tested for DNA. They all came down with personal stuff that belonged to the guys—toothbrushes, hairbrushes. Parents and children had to give saliva samples, so in case a body part was found, they'd have something to match it to.

"The fire station is still the fire station, but people aren't acting normal or saying normal things. We're working a lot, and we've had three good fires in the last four days. But we've got to finish burying these guys, too. That's what we have to do.

"It's a very strange situation for the firefighters who are coming in and taking these guys' spots. They feel very awkward. They don't know how other people are going to see it, or if it'll be like, 'Look who's slipping in.' It's a good house. It's a very busy, high-demand house, and it has a great reputation. Any other time, people would climb over somebody else's back to get in because it's a nice career move. So the new guys feel funny that they're coming in under these circumstances.

"I don't know what's going to happen to my head after this. They say, 'Cry a lot, work out a lot, sweat a lot.' The counselors come in and tell us what we're supposed to be feeling, what things to watch out for, things we can do to help deal with certain stressful situations. There are five different places we can go throughout the Boroughs and upstate that have counselors from the International Firefighters Union, the city firefighters union, social workers, volunteers, professionals, and doctors if you think you need medication. They tell us that we can tap out any time we want, that we don't have to worry about missing work or having it go on our record or getting points taken away if we're going for a promotion or transfer. They ask if we're sleeping, if we're dreaming, if we're talking, if we're crying, if we find ourselves driving and we don't know where we're going.

"Now everybody's asking me if I'm going to retire. They say, 'You've got your 20 years in, you're making a lot of overtime, it's a good time to leave.' But I say, 'I ain't leaving on this note. This ain't going to be the reason why I leave. It's going to be because it's time for me to go.' Then they say, 'You're never going to say it's time to go.' And that's true. I like going into buildings and putting out fires. That's what I like to do.

"I'm good at it. I grew up a typical tomboy, hanging out, liking things that girls aren't supposed to like. It excited me when I saw the fire trucks. I was always attracted to stuff like that. I like the rush, the adrenaline. I like the fact that I'm not afraid of the things I used to be nervous about when I first came on the job. I can go into a building and walk down a hallway with fire rolling over my head. That wasn't a very comforting feeling when I first got on the job, but now, I've done it hundreds of times.

"Even though it was rough in the beginning—you know guys, the way they can be—I actually liked the work. So I said, 'I'm not here to make friends. I'm here to work.' And it turned out that I ended up making friends anyway. And finally, after many years, the guys actually started saying good things out loud about me as opposed to just thinking them.

"Sometimes when I get detailed to another firehouse, I make sure I take my saltiest coat and my saltiest helmet and I just throw them down and say, 'Is the officer here? I want to see what my assignment is.' They look at me and look at my gear and look back and forth at each other, and there are usually no questions asked after that. If I had a brand-new turnout coat and a brand-new helmet, they'd be like, 'Where've you been hiding your whole career?'

"But that's what you've gotta do. If they talk to you, they talk to you. If they don't, don't force it. Don't let them know you're dying for conversation. I've gone to firehouses where no one spoke to me the entire time. I sat there, I ate, everybody talked, nobody addressed me, and no one asked me questions. It's like, 'Who's going to talk to her first?'

"So that's life in the FDNY. The women coming on now have it a hell of a lot easier. A lot of the old-timers who had their 20 years in when I came on are gone, and a lot of the guys who have 20 years in now have worked with a female and they're used to it. Somebody broke them in and made them realize that working with a woman is not as bad as they think it is. Actually, guys who were here when I first got here have daughters now who are old enough to apply, and they tell me, 'She's thinking about taking the fire department test.' They say, 'I gotta admit it, some women can do the job, right? I've worked with you for 15 years.'

"I've always said I'd love to make it through my career without losing anybody from my firehouse. Until now, we've never had a casualty. I hope no other fire department ever has to go through anything close to what we're going through, because we're all severely devastated by this—men, women, and civilians. It's not just the fire department.

"If they put buildings back up there, they'll have to have some kind of a memorial or a park where you can sit and have lunch. I think they should leave something original to the area. If they want to put buildings there, put buildings there. That would be a real kick in their ass. *Look. You knocked it down, but we built it up, and we're going on.* It's the World Trade Center. It's like the whole world in a little tiny area. You can't fold up and go home. You've got to make a statement and not forget the people who died.

"Life goes on, whether it goes on the same way or not. It's brought this city together. I've never seen people nicer to us or nicer to the cops. People who can't afford to put dinner on their own tables bring food to the firehouse. It's really great of them, but I'd rather be hated and have our guys back."

48
POLICE
DEPARTMENT
CITY OF
NEW YORK

MAUREEN BROWN

AGE 26
POLICE OFFICER
NEW YORK CITY POLICE DEPARTMENT
PROPERTY CLERKS DIVISION
QUEENS

Maureen Brown grew up in Queens in a family that has made civil service a way of life. Her father was a fire captain, and most of her siblings are firefighters or police officers. When the Twin Towers collapsed on September 11, Maureen lost a lot of people she knew in the police and fire service. That made her assignment at the morgue all the more difficult.

"Police officers and firefighters are all just people who put on a uniform," she said. "We have feelings, and we're everyday people. This has hit us all very hard."

Maureen said she's aged dramatically since that day. It was evident to us, too, as we listened to her story, that her young eyes have seen far too much. Although she expected to deal with death many times over in her career as a police officer, she said she never expected to see the things she's seen since September 11. Sharing her story with us left her emotionally exhausted.

"I know I need to get some counseling," she said, "and I'll do it in time, when I'm ready. But most of these cops won't ever go. This stuff will change them for the rest of their lives, and I think that's very sad."

"I've seen a lot of death. I've seen people stabbed, shot, decapitated, everything. But to see it like this in such mass—it was like war. Seeing someone's half scalp or someone's shoulder blade. I'll never get that out of my head."

—Maureen Brown, November 24, 2001

"As a female police officer, I know I'm always going to be viewed differently. This is a high-authority job. When I'm working with a male partner, I'm the one the prisoner or criminal is going to challenge. It's always going to be me. They think I'm the weakest one, so I'm the one they pick to go up against. Other cops may not want to work with a female, because they want another guy on their side who's going to be able to really throw the punches. But I'm a hard-working cop, and I've never had that problem. I always show what I can do.

"I've had to step up to the plate and prove myself. Every time I work with a different person, I have to think about what I'm doing, do it properly, and be strong. I want to show them, 'Listen. I'm a good cop. Even though I'm a female, I can probably do some things a lot better than you.'

"I think women police officers do a good job of assessing situations, and there are certain jobs that men sometimes don't handle as well as women do. When it comes to dealing with children or victims of rape, they usually want a female there because it's a little more sensitive. Women can bring themselves to the job on a different level.

"I've been a cop for four years. I was working patrol in the 48th Precinct in the South Bronx, and I was arresting someone, rolling around with a prisoner, when he resisted violently and I got hurt. A couple of other cops got hurt, too. I tore the cartilage off my shoulder and had to have surgery to put it back, but it didn't take. I'm doing physical therapy three times a week, and I'll need to have surgery again.

"After I got hurt on the job, I was put on restricted duty and transferred over to the Property Clerks Division in Queens, where we take in the property of deceased people, prisoners, and anything else that is found or seized. Any kind of property that ends up coming into the hands of the police, including drugs, money, guns, and narcotics, is vouchered and stored here for safekeeping.

"After the first plane hit on the morning of September 11, we were watching it on TV at work, and it was like a crazy madhouse. Everybody was in a state of shock, and we all felt helpless because we couldn't leave, and there was nothing we could do to help. We have all the drugs and guns at the Property Division, and we were on high security alert. The Property Division is a huge warehouse where we've stored everything that's been taken off the street for years and years, so we have thousands and thousands and thousands of guns. Anybody could come in here and try to take over.

"I'm not really supposed to be in uniform or out on the street because of my shoulder, but that day, we all put our uniforms on and were ready to take action. People didn't know what to do, and all the phone lines were ringing. Then all of a sudden, the phones got shut down. None of the phones were working in the whole area, because the Tower with the big antennae on it had just gone down. We stayed there that night until very late, and then we were told, 'All right, you guys have to go down to work at the morgue tomorrow at 6 o'clock in the morning.'

"When I first came to the Property Clerks Division, I didn't realize that this was part of the job. As a cop on the street working regular patrol, I'd take certain property off a prisoner and voucher it for safekeeping. If a person passed away, I'd go in, secure the area, and if there was something of value within eyesight, I'd voucher that so it wouldn't be stolen once the body was removed. I knew all that stuff went somewhere to be organized and controlled and kept safe, but I didn't think much about it. I didn't know that when there's a major incident, the people in the Property Clerks Division go down there to voucher everything and take control of all the property.

"I thought, *Okay, we're going down to the morgue, and that's our job for the day.* I didn't realize how bad it was going to be. When we got down there, the whole street was blocked off, and huge refrigerated trucks, like meat trucks, were parked on the side street. I didn't get it at first that the reason those meat trucks were there was for the bodies. They were expecting all these bodies to come in, and they had to store them, obviously, in a freezer, or something along those lines, where it's cool, so the bodies wouldn't get bad.

"We saw one of the guys from Property who'd worked there that night, and when he came out, he was crying. You know, he's a grown man, and he's crying. It was so moving for me to see a guy cry, a police officer. He said, 'I'm telling you right now, you need to brace yourself. You're going to see a lot of dead firefighters, and you're going to see a lot of dead cops.'

"At that point, they were mostly finding the bodies of firefighters, because they were farther out in their trucks when the buildings came down. Not only that, but they were the only ones who could really be identified, because of the fireproof jackets with their names on the back. Identifying civilians was a lot harder, because their bodies were decapitated or burned or destroyed, and there was no way to tell who they were. On the news, they would give a body count of how many people they had found, and it was mostly firefighters. Firefighter after firefighter.

"One of my brothers and my two brothers-in-law are firefighters, and my father was a fire captain who died in the line of duty. One of my sisters is a police lieutenant, and my other brother is an Amtrak police sergeant, so we knew a lot of people who were missing. Everybody knew I was at the morgue, and they'd call me up and say, 'This person's missing, and that person's missing. Let me know if something comes in.' So every time I came in, I'd look at the paperwork from the night before to see if they'd found that person.

"One of the first people who died was Father Judge, the fire department chaplain, who was close to my family. He'd been hit by debris while he was giving last rites to one of the firefighters who died when someone jumped out of the building and landed on him. Father Judge married my sister and brother-in-law, and he'd known my brother-in-law's family for years. One of my brothers talked to him almost on a daily basis. When I got to the morgue, they'd just removed his body, thank God, because I don't know if I could have handled that. His death was devastating for my family.

"The morgue was at Bellevue Hospital, and it was somewhat of an assembly line because of the number of bodies and body parts coming in. A small truck would come through the barriers, and you'd know that these were new bodies coming in. During the first few days, the bodies were coming in very quickly. They were finding 600 body parts a day—arms, legs, scalps. It was just horrendous stuff to see.

"The detectives would pull a body bag off the truck and walk it through the process so that they wouldn't lose track of it or mix it up. All the autopsies were being done on one side of the building, and there were several doctors in there. Then the bag would come around to where we were, and we would voucher the property. We had long tables, and the detective would unzip the body bag, go through it again, and say, 'All right, voucher this, this, and this.'

"At the first table were the people who would take the property and put it in a plastic bag. Then we would pass the body bag down the line, and the doctor would say, 'Looks like a piece of scalp, possibly male.' Then we would have to write this down, because each body or body part would get a medical examiner's number. So we wrote the number down, and if the doctor said, 'Male torso,' we wrote down 'Male torso. Nothing found. No property.' Or we'd write, 'Female hand, wedding ring found. Ring says, "Forever, Shawn"' or something. Horrible things like that.

"Eventually, they will do DNA testing on the hand, and hopefully they'll be able to identify that person. Then the wedding ring that was taken off the hand will be matched up with the ME number, and the property will be returned to the family.

"After we vouchered the property, the detective would take the bag back to the meat trucks. The trucks had an American flag hung over each one of them, which I thought was pretty nice. Any time they found a cop or a firefighter or an emergency worker, they wouldn't bring it in on the truck with all the other bodies and body parts. They'd put it in an FDNY ambulance and bring it in like that, by itself, out of respect.

"The first day I was down there, we vouchered the property of Yamel Merino, a female EMT from MetroCare. She had a cell phone, and the girl who closed up the bag was supposed to remove the battery, but she never did. So Yamel's cell phone was constantly ringing, and we felt horrible. Once it's sealed in a bag, it's sealed in a bag. You can't open the bag, because everything gets a number and it has to be in a certain order. So her phone was ringing and ringing and ringing. The person from MetroCare came to pick up her equipment, and the guy said that Yamel was a single mom with an eight-year-old son. That made me feel even worse. When I heard stories about people's lives, it made it more personal. It was a lot easier to remove myself when I didn't know the details.

"One firefighter they brought in was so badly burned that his helmet was caved into his skull. It was a terrible thing for me to see. My dad was a firefighter, and he died in the line of duty in 1982, so it brought back a lot of feelings for me. It was the last place in the world I wanted to be. Seeing all these dead firefighters being brought in was just so sad. Last Father's Day, there was a big fire in Astoria and three firefighters were lost. It's very hard for the city to bury one or two firefighters a year, let alone 343. More firefighters were lost on September 11 than have been lost in the whole history of the fire department.

"My heart really goes out to all the firefighters who lost members of their companies. I can't even imagine how they can still go to work. The firehouse around the block from my house lost 19 people, and now they have all these wives and children sitting there, waiting there, day after day in the firehouse, waiting and waiting. Every time the firefighters came into the morgue to take another body out, they'd be crying.

"One of the things the rescuers found was an arm with a tattoo that said 'FDNY,' so we knew it was a firefighter. They also found a stewardess's torso and hands, and the hands were tied together, like they'd been tied behind her back. No one is ready to see something like that.

"I was doing 14 hours a day down there, and the smell was horrible. The smell of death is horrible in itself, but with the massive amount of body parts they were bringing in and the burn victims, it was such a horrible smell. I could smell it on my clothes, and when I went home, I'd take a shower and scrub myself down, but no matter what I did, I could still smell that smell. I will never, never, never get that out of my head.

"Everybody was silent down at the morgue. It was totally dead silence. People were like zombies. We were working 14 hours a day, we'd go home, we'd watch the Towers coming down over and over again on TV, and we didn't sleep for weeks. Every now and then, someone would break down hysterical crying. Guys were crying, girls, everybody. I didn't really cry until two days into it, because I was in such a state of shock. But to see guys cry, and cops and firefighters, people crying all over the place, all these higher-up captains and stuff, it made me think, *What did they do to us? What did they do to our city?*

"They were rolling in one body after another, and somebody working there would suddenly say, 'I know this person. I went to school with this person.' We didn't know who they were going to bring in next. I remember one girl who was down there when they brought in somebody she knew and how bad that was for her. We all knew people who worked in Lower Manhattan or who responded to the World Trade Center, and at some point we were all going to see someone we knew. We were going to voucher someone's driver's license, someone's wedding band, someone's jewelry, and it would hit us hard.

"But we had to do it. Who else were they going to have do it? You're not going to have all these out-of-state people do it, like the people from the FBI or FEMA. This is New York, and we have to take charge of it ourselves.

"It was so bad down there that every now and again, someone would just break down and have to leave. I'd have to try to compose myself, because I had to be professional about it in order to get the job done. I'd go home and talk to my sister, and I'd cry to her and say, 'This is so horrible. I don't know if I can do this every day.' Then I'd talk to my brother, who is 30-something years old, and he'd be crying to me because he lost all his friends. He'd say, 'No, Maureen, you have to do this job. You have to be the one who's down there doing this. You have to do it, because you'll do it the right way.'

"I understood what he meant by that, because not everybody could do this job. My brother knows I care. He knows this hit home for me and my family, and he knows I'm going to make sure that this person's property gets vouchered the right way. This wife is going to get her husband's wallet and that husband is going to get his wife's wedding ring, because that's closure for the family, and that's important to me. That will help people in the long run. I said to my brother, 'You're right. You're right. I'll do it the right way, and I'll do it respectfully for the families who have lost these loved ones.' That kept me going.

"I was having nightmares. Terrible, terrible nightmares. I don't know if it was because of what I'd seen, or just because of the reality of what was going on in New York City. As a police officer, I've always felt very secure and very confident. September 11 knocked that right out of me. It's very sad the way people are living now in a state of fear. Like, *What will happen next?* No one thought this would ever happen. Not in a million years. And now, every day, you have to think about what's going to happen next. *Are my children going to grow up?* It's a terrible life now, living in fear.

"Any time there's something major going on in the city, my family is always there. We're all in civil service, so there's such a high chance of one of us getting killed. Fortunately, we didn't lose anyone on September 11. My brother always jokes that, statistically, because we already lost our father in the line of duty, the rest of us should be okay. But you know, anytime something big goes on in the city, the whole family is there. Everybody's working, everybody's affected.

"My brother lost one of his closest friends in the Father's Day fire, and now this. Thirty of his good friends are gone. My sister is a lieutenant in the 40th Precinct, and they lost Officer John Perry. He was putting in paperwork for his retirement at headquarters when this incident happened, and one of the captains grabbed him and said, 'Let's go down there.' They got into this big fireball inside one of the buildings, and the captain saw a firefighter's light and followed it out. He thought Officer Perry was behind him, but the building came down and Officer Perry was killed.

"It's all still so shocking that I just try and push it away. Any time I turn on the TV there's something on about this, and it brings it all right back. I try not to watch TV. I try not to read the paper as much, because I don't want to know any more. The NYPD is still on Omega status, the highest police alert we've ever had. We continue to work and keep going. We had another plane crash out in Queens a few months after September 11, and now we're vouchering all that stuff, taking in all that property, and it feels like it just never ends. It probably won't end for a long time.

"When I close my eyes, I see things. When I feel like it's taking me over, which it does sometimes, I try to think of other things that are important to me, like my family. If it takes me over, I know it will destroy me. I would be in such a state of depression, and I'd be miserable. I know that could happen. I know I need therapy. I'm sure as hell going to get some, but I need to do it in my own time. I know if I go in there and start releasing stuff now I'll shut down, and I won't be able to deal with my job. I have a job to do, and that's how I look at things right now. I have a job to do, and until my job gets done, I've got to take care of everybody else before I can take care of myself.

"There are a lot of firefighters who say, 'You know what? I lost a lot of brothers, and I'm going to take this job and do the best I can with it, because that's what they would have wanted. They would have wanted us to move on and be strong and not let the terrorists get to us.' But I don't know. I want out. I know a lot of people who want out. I know so many cops and firefighters who just don't want this job anymore. They love their jobs, but how can you continue working when you've lost all your friends?

"I took this job to help people, but I didn't take this job to see what I've had to see. I can deal with death, I can deal with helping people and saving people, but to put myself in a situation like this—no one is ever ready to deal with that. No one. No matter what line of work you're in, even going into military, you don't sign up for this. You go in because you want to protect your country or you want to protect people. As a police officer, I knew I would see death. I went to the morgue during my training, and they showed us all this stuff, but you don't know until you actually see it how it's going to affect you.

"I remember the first time I saw a dead body. It was probably two days after I got on the job, my first time out on the street, and this guy hung himself. He was hanging over a banister. After the first couple of weeks on the job, I started having nightmares because I saw so much, working in such a bad area, the South Bronx. But I liked it because I was active and I was busy and every day was different and exciting. I could do that, and I could remove myself from the situation and deal with it.

"But when something happens like this, I can't deal with it. No one can. But this is my job now, and I have to do it. I'm ordered to go down there, and I have to go down there. I have to work at the morgue, and I have to do this, this, and this, and I can't say no. I have to do it because who's going to do it if I don't?

"I didn't sleep for a month, I swear, because how could I close my eyes after what I'd seen? Knowing what was going on in the world, I didn't feel secure. Being a police officer, I'm supposed to be in charge of things, take care of things, fix things. But I can't. I can't do anything because I'm devastated myself.

"I was down there for three days the first week and then I was removed. I think it was because it was getting really bad, and they realized that I knew a lot of people who had passed away. The sergeant said, 'You know, we only need a certain number of people down here. Why don't you go back and do security for the Property Division?' Maybe it was because I've only had four years on the job and he didn't want me to be seeing all that stuff. Whatever. You've seen three days of it, you've seen it all.

"As soon as all this stuff started going on, I wanted to go home. I wanted to be with my family. I'm a cop, and I have a job to do, but you know what? I'm a person, too. I wanted to be home. If something's happening and this city's getting blown up or something major is going on, where do I want to be? I want to be with my family. But because I'm a cop, I have to go out there and go to work while everybody else is going home to be with their families. That's hard. It's hard to do.

"That day, when everything was happening, my sister was crying for me to come home. She was there by herself with the kids, and I felt bad for her. Her husband's a firefighter, and he was out on the job. Everyone was working except her, and she was home watching all the kids. I told her, 'I can't come home,' even though that's what I wanted to do. I'm a person, I'm a human being, and I have the same feelings as everyone else—but once I put this uniform on, I have a job to do. That's pretty much what I did, and it's what I continue to do.

"I've seen a lot of death. I've seen people stabbed, shot, decapitated, everything. But to see it like this in such mass—it was like war. Going through a guy's wallet to voucher money or credit cards and seeing a picture of his little kids. Seeing someone's half scalp or someone's shoulder blade. I'll never get that out of my head. I hope nobody else will ever have to see anything like that. I would never wish this on anyone."

MAJOR
MOLLY SHOTZBERGER

AGE 59
COUNSELING COORDINATOR
FOR THE WORLD TRADE CENTER DISASTER
SALVATION ARMY
NEW YORK CITY

Major Molly Shotzberger told us that one of the women she counseled at Ground Zero likened having the Salvation Army onsite to having "Mom" at the ready. Molly, herself, made us feel like we were sitting at the table with Mom as she looked deep into our eyes and asked how we were dealing with hearing all the stories we were collecting for this book.

Although Molly's mission was to bring comfort and a calming presence to rescue workers at Ground Zero and those working at the morgue, she never pretended to have all the answers.

"I didn't try and fool people into thinking I could tell them why this happened," she said. "I don't know why it happened. All I know is that it did happen, and we have to go on from here."

Molly and her husband have two grown children and four grandchildren, and her family is proud of the work she is doing to help rescue workers and grieving families at Ground Zero. Even as her grandchildren share stories about her experiences with their classmates at school, Molly is reluctant to accept any praise.

"It's the rescue workers they should really be proud of," she told us. "It's the families of the victims they should be praising for their strength."

"We would take their boots off, take their socks off, dry their feet, powder their feet, massage their feet, and put on clean socks. As we did this, they would begin to talk."

—Major Molly Shotzberger, November 25, 2001

"Many people think the Salvation Army is a social service organization. However, even above and beyond that, we are a religious organization. We are a church, and all our social services are an outreach. We run the gamut of social services. Our motto is, 'Heart to God and Hand to Man.' We believe that because God has ordained us as ministers, it's our responsibility to go out and show His love in practical ways, more than just preaching the Word.

"My husband and I are both ordained pastors, and we share a dual responsibility. We run the company store for our territorial headquarters, which means that we provide all the materials for the Salvation Army to do its ministry. It's really not a job. It's a spiritual calling, which is why, in the Salvation Army, both the husband and wife are ordained. If you both share that same sense of calling, then, for instance, during this September 11 crisis, when I was gone day after day and working 18-hour days, my husband shared that same commitment. He was my biggest supporter, in that he released me to go and do what he knew my passion really was.

"How I got involved with September 11 was that I was part of the counseling team at the TWA Flight 800 crash. That was five years ago out in Long Island. Being stationed in the division where it happened, I was sent to the medical examiner's office to do counseling with those who were identifying the bodies. During that experience, I realized how inadequate my training had been, and since then it's become my passion to get better qualified, better trained, in case something ever happens again. We service fires, floods, and those kinds of things all the time, but September 11 is the most tragic thing we've ever been involved in.

"I was late going to the office that day. I turned the news on as I was getting ready to leave the house, and the first plane had already hit. I was on the phone, calling my husband, saying, 'You're not going to believe what just happened,' and as I was telling him this, I watched the second plane hit. I just started to cry.

"About half an hour later, I got the call to go to the medical examiner's office and set up a counseling team there. I took a number of Salvation Army officers with me, both men and women, and talked to the head personnel over there. I introduced myself and told him of my experience with the Flight 800 crash, and they welcomed us with open arms. This was during the first couple of hours, and everybody was in such a state of shock. We were trying to deal with our own grief and still help those who were eventually going to be trying to identify the remains.

"I left that team at the morgue and went over to Ground Zero to assess the situation, and then called in another counseling team to go down there. At that point, the Towers had fallen. I was thinking that all those people who died were probably never ever going to be found, and how horrific that was. All I could do was cry, and I was thinking, *How do these people feel who actually had somebody die?* I mean, many people lost a husband, a wife, a brother, a sister, a father, a mother, or a child. It was so big that I just couldn't grasp it. How was I going to help these people? I just prayed, *God, give me the words to say to them.*

"We first set up in the American Express building, where they were bringing the bodies. We didn't have food, but we had snacks, water, Gatorade, those kinds of things. It soon became evident that there were going to be a number of body bags coming through there, so they felt that because of contamination, it was probably a good idea that we move outside. In the process of trying to get established outside, we realized that those who were out there digging in the rubble were totally exhausted. You know about the heat, and how hot it was, and how dehydrated they got, and they just wouldn't stop because, in those early days, everybody was just hoping to find one live person.

"We realized they weren't going to come to us, but they needed what we had, so we crawled through the back of the building, carried over water, and literally started a bucket brigade. We filled buckets with ice and Gatorade and chocolate bars, because that's what they were asking for. We had a bucket brigade right up on the rubble to those workmen and -women. We were passing up the cold drinks, chocolate bars, and anything else they needed.

"We also set up a medical tent for their eyes. We started washing out their eyes, and we noticed that the fingers of their gloves were all torn and their fingers were bleeding. So we helped them take their gloves off, bathed their hands in saline, and then put salve and bandages on where we could. There were a lot of new gloves that had been donated, heavy work gloves, so we would put those back on their hands.

"Then we realized, when they came out of the rubble, that their feet were really, really tired. The soles of some of their boots had burned through and their socks were just soaked. So, when we could get them to sit down, we would take their boots off, take their socks off, dry their feet, powder their feet, massage their feet, and put on clean socks. As we did this, they would begin to talk, and they would begin to relieve some of their stress. Then we'd put their boots back on and send them back in, wishing there was more we could do.

"The women working down there were wonderful. Those women were just as tough as any of the men. They worked so hard. They were unrelenting. I think because there were so many more men than there were women, the women got overlooked. But from my perspective, I want to tell you there were some real heroes down there. I don't know their names, but boy, I saw them work. I saw them in there digging as hard as the men. I saw them reluctant to take breaks, just like the men, and when they did, there was perhaps more urgency in the women to get back in there and get started again. Perhaps it's that female instinct. They'd say, 'Well, I'll have a cold drink, but I've got to get back in there,' even though they were exhausted.

"They were just hoping they could find somebody alive. That's what kept them all going. There were a lot of tears. There was anger that this had happened, that we had been so vulnerable, that there wasn't something that could have been done to stop it. There were some who wanted to talk about what they were seeing and others who were just so sick of it that they didn't want to talk at all. There were those who could share their feelings very easily and those who were protecting themselves. Some seemed to be afraid that if they let the first tear come, they wouldn't be able to stop. Others were absolutely numb and didn't know what they felt.

"Of course, there was survivor guilt. People would say, 'I know my friend is in there, probably dead. Why did I survive?' Then there were the rhetorical questions. 'Why did this happen? Why did God let this happen?' I had to be honest with people and say, 'I don't know.' People were angry with God, and we tried to reinforce that it was okay to feel that way. They seemed surprised to hear a minister say, 'It's okay to be mad at God.' Their emotions were running the gamut, and we told them, 'Whatever you're feeling, it's really okay.'

"A lot of people had difficulty trying to relate to their families. They'd say, 'How do I explain this to my wife?' or 'How do I tell my children what I'm doing down here?' Our advice was to not be very graphic. We said, 'Just tell them you're down here trying to help the families, trying to bring some sense of peace to another family.' And they would say, 'Well, my wife is reacting like this,' or 'My child is doing that. Is this normal? Is it okay?' And they would tell us about the nightmares that many of their children were experiencing.

"On my second day down there, some of the buildings were in danger of falling, and nobody was sure how stable they were. We'd been told, 'If somebody yells "run," you don't look back, you don't ask questions. You run and you get out of there as fast as you can.' I was walking up to the pile of rubble, and all of a sudden I saw these men and women come running out yelling, 'Run! Run!' I had a laminated ID tag around my neck, and I remember putting it in the palms of my hands and clasping them together as I was running, thinking that if I were to die, maybe somebody would find my ID tag and be able to identify me. Then, as I ran, I had such an incredible peace come over me. I looked heavenward and said, 'Okay God, if this is it, this is it. I'm ready.' And from that moment on, I was not afraid to go back.

"We had a nurse who teamed up with us for the first few days. She came down on her own, wanting to do something. She said, 'Look, I want to be helpful,' so we said to her, 'Fine. Be one of us.' That's how we started reaching out to people who lived in the area. They were forced out of their homes and had no place to go, and until they started the ID process, they were allowed in the area. They became some of our best volunteers.

"I remember one woman in particular who lived nearby, and by staying with us, she could at least watch her apartment. She had gone over to the canteen to get something to eat, and when she came back, she was absolutely frantic. I said to her, 'What's the matter?' And she said, 'My umbrella! My umbrella!' I mean, she was shaking. I said, 'Well, where is it?' And she said, 'I don't know! I lost it! I had it right here!' I said, 'All right. Let's take a deep breath and see if we can find it.'

"Well, I found it lying on the table, and I want to tell you, when I handed her that umbrella, her whole demeanor changed. She let out a huge sigh of relief. It turned out that this umbrella was the only thing she had taken from her apartment. It was her only connection to anything she owned, to her life before the attacks. That taught me a lesson. In the midst of this horrific event, a person who has lost an umbrella can be in just as much crisis as anyone else.

"We had a guy named Sam who had been walking his dog near the Towers when they collapsed, and in running for his own life, he dropped the leash. Of course, the dog got lost, and every day he'd come by the tent. 'Haven't seen my dog, have you?' 'No, Sam, we haven't seen the dog, but come on over and have a cup of coffee. Have a cold drink, a chocolate bar.' I learned a lot about those 'incidental' crises that in the midst of everything else could have been forgotten. So while we had this huge, gigantic, mammoth tragedy, there were all these other personal tragedies going on that people didn't see.

"In the beginning, the Salvation Army teams were from West Nyack, but after the first week, they came from Syracuse, Rochester, Albany, Buffalo, and Jamestown, New York, as well as Ohio, Kentucky, and all over the United States. We had teams come in from other territories to staff the canteens, and they were providing food as well as medical supplies. I even had one firefighter from Nova Scotia call me, wanting to do something. He said, 'I just feel so helpless. My motorcycle's packed, and I'm just waiting for your call.'

"As I brought other teams in, we had medical personnel, doctors, psychologists, and psychiatrists. I told the mental health people to throw away the textbooks.

I said, 'We're here to meet the need at the point of need. We're not here to analyze their whole lives. We're not here to deal with the problems they deal with every day. We're here to get them through the moment.' For some of them, in the beginning, that was a little hard, because they were trained to do just the opposite. But it worked. I said to the team, 'Our mission, above anything else, is to bring a calming presence to these situations.' Whether it was at the medical examiner's office or down at Ground Zero, if people could see a calmness in us, then hopefully we would be able to pass that on to them.

"For the first month, I was between Ground Zero and the medical examiner's office every day, running the counseling teams. I really felt that I needed to keep my hands in it to know how my teams were doing and how the folks we were trying to help were doing. In a situation like this, it's so easy for it to become a 'job' rather than a 'mission,' and I didn't want that to happen. In a job, I would have gotten the teams organized and gone back to the office to run things from there. But if it's a mission, I'm part of it. I'm not removed from it. It was a team effort, and I wanted to be there.

"When we first set up at the medical examiner's office, I went around and introduced myself to people, and I told them we were there if they wanted to talk or if they needed anything. On the second day, some of the corrections officers came up to us and said that they needed some flags. So we called our Divisional Headquarters in Manhattan and said, 'Bring us the biggest flags you can find.' Within an hour, they brought a number of flags up, and we didn't think any more of it.

"A little bit later, the officers came over to us and said, 'Remember the flags?' They pointed down the side road, and there were three or four big refrigerated trucks that had been draped with those American flags. They were putting the bodies in there, in the refrigerated trucks, and with tears in their eyes they said, 'We're going to give them some dignity.'

"When they brought in the body of a rescue worker, there was a regular routine. The ambulance would come down First Street to the medical examiner's office, which is part of Bellevue Hospital. It would have a police escort, and everybody would stand at attention on both sides of the road. We would line up, and when they brought the body bag out of the ambulance, there would be an American flag draped over it. And as they lifted the body bag with the flag, we would all salute. Then they would take the flag off and fold it like they would at a military service, and the captain would say, 'As you were,' and the salute came down. Some of the people walked away then, but the counseling team would move among those who were still standing there. Sometimes during that very solemn ceremony, somebody would just reach over and squeeze your hand, or you'd see a tear come down and you'd pass a tissue.

"I remember one woman from the Corrections Department who had witnessed this for the first time, and it was very emotional for her. I took her aside, and we just talked.

She was feeling so sad, and she needed to be able to talk about it. A lot of it was listening. You know, many people have this mistaken idea that counseling is talking, but a lot of it is listening, letting people vent, and reassuring them that whatever they're feeling, it's okay. It's okay.

"We had an opportunity to pray with a number of people as well. We prayed with them for strength to be able to go back in, prayed with them for protection, prayed with them that their families would understand why they had to be away from home. Some of our folks were actually called upon to pray over remains when they were found and lead the procession down to place the body bag in the ambulance. That's what religion is. Religion isn't something in a book. Religion is what you do for other people. Christ said, 'I was hungry, and he gave me something to eat. I was thirsty, and he gave me something to drink.' That's what we do in His name.

"Yes, I had tears, too, but when I really cried hard was in the car on the way home. I felt that if I fell apart in front of those people I was trying to help, I wouldn't be able to be very helpful. It was okay to cry with them and to shed tears, but to cry uncontrollably was another thing. I hurt, just like they hurt. I was sad, just like they were sad. I was angry, too. I'm human, just like they are. But I've got an incredible faith that helped me deal with all of this and get through each day without breaking down. So I'd cry in the car or while watching television at home.

"My husband stayed at the office and kept things going there. But he was so supportive of me. I'd come home and be so exhausted, and one night I remember saying to him, 'Oh honey, I think I'd like a hamburger.' Well, he not only made me a hamburger, he actually cut up lettuce and tomatoes and made me a little salad. I'm eating this, and then he comes in with a plate of cheese and crackers and pepperoni. I was so touched that he went out of his way to make things nice for me at home that I cried all over again.

"Down at Ground Zero, I felt that there was no race, no gender, no religious separation. We were all one. We were each doing something different, but we were all working for the same purpose of trying to bring peace to the families of those who were lost. Whether it was digging for survivors, identifying remains, or taking care of each other, there were no differences among us.

"I think this whole experience has changed me in a number of ways. First of all, my faith has gotten so much stronger because of this. I will never again look at the American flag without a tear in my eye. I will never again look at rescue workers without a whole lot of pride in my heart. I will never take my freedoms for granted. I will never forget the unique ways the American people found to be helpful, to make it easier for those of us who were down there. And I will never forget the families."

534

LIEUTENANT ELLA McNAIR

AGE 43
LIEUTENANT
FIRE DEPARTMENT OF NEW YORK
ENGINE 283
BROWNSVILLE, BROOKLYN

Ella McNair invited us to her comfortable home in Brooklyn to talk about her career in the fire service and the events of September 11. Although she was warm, welcoming, and willing to talk about her experiences, Ella's interview was considerably different from the rest. She guarded her emotions carefully and told us that she was reluctant to share her feelings too deeply.

"I haven't really let myself feel it yet," she said of the tragedy. "I keep myself busy. I have my moments when I reflect back on what happened, but I don't revisit it much in my mind. I've been down to the site quite a few times since then, but I don't want to stay there mentally. I have to look forward. Things happened for whatever reason, and I hate that they happened. But for me, it's not good to stay stuck."

One of the first women to enter the Fire Department of New York in 1982, Ella has worked in many different firehouses. A few days after our interview, she was promoted to lieutenant.

"I still can't really comprehend the magnitude of what happened because I don't want to. I just leave it like that. I was there, I saw some things, but I don't go into it too deep."

—Lieutenant Ella McNair, January 27, 2002

"I was home when I got the call, and they told me to come into work. The firefighter who called said a plane just went into the Trade Center. I thought he was joking at first, but after he said it again, I wasn't taking it as a joke. As I was going in, I saw other firefighters who work in the area, and we were passing each other in our cars. At that point, I realized that maybe something big was going on.

"I'm not far from my job, maybe 10, 15 minutes at the most. My firehouse was the gathering point for the division, and when I got out there, not many had gathered yet. They told me we were all going right out to Manhattan. The few of us who were there got on a spare rig, went to pick up some radios from another house in the area, and by the time we got back to my company, the whole place was packed with firefighters. I was lucky I was on the rig. They had to wait for a bus.

"When we were going across the bridge, I wasn't on the side to see the World Trade Center. I really couldn't see what was going on. One of the guys was saying that one of the buildings was coming down, but I didn't really believe him. We say a lot of things in the firehouse, so you don't really take everything seriously.

"When we got over there, I saw all this dust on the street. It was like total chaos at first, because nobody knew what to do. It was like havoc over there. I was just in disbelief. I didn't want to think that these guys were in the buildings. And then I was thinking it could have been me when all this came down. So many different things were going through my mind. I didn't want to believe what I was seeing.

"The first Command Post was lost right away when the buildings came down, and that's where most of the chiefs were killed—Bill Feehan, Peter Ganci. We finally found someone in command, and he told us to go and feed some water from the river. But that didn't work out because we were waiting for a fireboat to come to hook up with, but it never showed up.

"We wound up going over to where one of the buildings had been, and we started digging. We were looking for survivors. I didn't think we were going to find anybody there. I was looking, but just seeing the magnitude of what had happened, it was hard to comprehend that anybody could survive this. If anybody could have survived, they would have been in the voids, and we should have been able to hear them. My company didn't have any cameras to look through, imaging cameras or anything like that, so we just went by feel, looking through voids, and calling out.

"At one point, they were telling us to back up, back up, and we ran, and when we ran out of the way, another building came down. It was one of the hotels. I didn't really see it, but I listened to them, and when they told me to get out, I ran. I didn't even look back.

"Later, we went over to another section where some guys were trapped. We thought they were guys who got trapped when the buildings came down, but it turned out that they were looking for other guys and they got trapped themselves. We had to feed them oxygen because it was really bad. It was very dusty and smoky down there. They kept saying, 'Get air tanks! Get air tanks!' So I was running back to Engine 10, which was right there in front of the World Trade Center, and getting masks and air tanks, bringing them up and passing them over the rubble to wherever they needed it.

"I was hoping other people would come out of it, but with all those floors coming down, I couldn't see it happening. I know the firefighters weren't on the top floors. They probably were between the bottom floors and middle floors somewhere, so I just couldn't see how they could have still been alive.

"When I'd see some of the brothers and sisters from the fire department, I was happy to see them, because I didn't know who was missing at that time. I was just so happy to see somebody I knew. It took a couple days before we really started getting a count of approximately how many people were missing. The atmosphere was different as far as how we felt about each other that day. Even back at the fire station days later, when I saw somebody, it was good to see him. You know, I'd hug him and say, 'It's good to see you.'

"We got to Ground Zero about 10 in the morning, and we left there about 12:30 at night. I went back three days that first week. We were just digging, looking for any kind of body parts or whatever we could find that was human. I didn't like being out there. I really didn't. All kinds of things started looking like body parts. Even at night, everything looked like it was part of a body. Some people wanted to be out there, but it wasn't a good feeling for me. I felt there was no hope in getting anybody out of this, but I was there, digging anyway. Some things you just have to do.

"I still can't really comprehend the magnitude of what happened because I don't want to. I just leave it like that. When I think about it, it still makes me wonder what the hell really happened over there. I was there, I saw some things, but I don't go into it too deep. I know what happened; I know how many lives were lost. A lot of them were my friends.

"First Deputy Commissioner Bill Feehan was a good friend of mine. I remember a couple weeks before this happened, he was saying, 'Ella, before I leave this job I'm going to see you promoted. I'm going to be at your promotion.' He was a good friend of mine, and now that he's gone, he's not around to see me get promoted. I'm not a religious person, but I have a belief that there are spirits who guide me. I feel that Chief Feehan will still be there when I get promoted. His spirit will be there. And others, like my parents who passed on, and all the people I lost at the World Trade Center, their spirits are going to push me through the challenges of this job. The wind beneath my wings, they are.

"We lost one guy from my house, Vinnie Morello, who'd been there just a year. He got rotated to Manhattan a couple months ago. And then there's another guy, Jimmy Gray, who wasn't assigned to our house, but he rotated to our house for a year. He was one of the guys who was caught up in it.

"I have a spirit in my life. Not everybody believes in the spirit I believe in, God or whatever. That's what keeps me above all this and a lot of other things I might feel I can't handle.

"I'm sad for the people who lost their lives. I hear all these stories, like a mother who was walking with her child and the child got lost at the time. I get a lot of stories. I just feel sad that we lost so many firefighters. I know that's part of our job, going in there, and we would do it again today if we had to. I would do it. Nobody ever thought the buildings would come down. This was not a normal thing, so, you know, you can't blame any of the firefighters for going in there. If something happened tomorrow at the Empire State Building, I would go in.

"I don't know why it happened. To me, there are lessons to be learned in everything in life, and I'm looking to see what the lesson is behind this. I'm still waiting to see what comes out of it. I don't want to believe it happened, so I'm not even looking for a lesson yet. I'm not at that point.

"I don't have nightmares about it. I just think about the people who were close to me who went on, and I wish that it wasn't them. I don't know how many funerals and memorial services I've been to because I haven't been counting. I just go. There are just so many guys who have passed. I try to be there when I can and show my support to the families.

"A lot of people have been showing up from neighboring fire departments, from the next town or next state over and places far away. You know, people are just showing up for support. Last month, the state of Louisiana brought a fire truck to our house and donated it to my company. That's how much people are pouring themselves out. We lost a lot of equipment. I think 90-some pieces of apparatus were destroyed.

"Right now, it's a good feeling to be a New York firefighter because people recognize that what we do is a special thing. It feels good that people recognize our service to the public and that they're showing support and appreciation for what we do to help them.

"Before I became a firefighter, I worked at Merrill Lynch. It was a desk job, all day, all day, all day. When I heard that the fire department was accepting applications from women for the first time, I wasn't really interested, especially because of all the physical stuff I'd have to do. I said, 'I ain't interested in that,' but they just kept sending me letters.

"At the time, I lived on a block that had five firemen, so after I got tons of letters, I went to one of the guys and he said, 'Girl, go for it.' I said, 'Okay,' and I went for it. I got a trainer, and we trained every day, and I felt like I was on top of the world. I felt very confident as far as being able to pass the physical exam, and I did.

"The Fire Academy was good. It was nice to be in a girls' locker room and have the camaraderie of being in a group, the females together. I didn't mind it when they yelled at us in the Academy because I didn't take it for real. I knew they wanted to yell down my throat, and it was okay. It was like, 'Okay, that's your thing, do your thing,' you know. I don't want to talk about the first firehouse I went to. It wasn't good. I was up for the challenge, though.

"I've been in the fire service since 1982, and on Tuesday, I'm being promoted to lieutenant. That means I'll be in a supervisory position, where I'll more or less delegate the work to the guys. It will be my job to make sure that everything goes smoothly and make sure all members of the crew carry out their jobs. We work together as a team. What one does helps the others. I'll be in a position to supervise and coordinate the work that goes on. I won't be just standing back. I'll be in there searching or doing whatever I can to help my team.

"I love my job. I love the adrenaline. I love going in there and being able to say, 'I put that fire out.' It's a good feeling. I love working with the guys. In the firehouse, it's like a family atmosphere. Sometimes our personalities clash, but it's all a family, and we eventually come back together. I'm a very adventurous person, and this job offers me an opportunity to do what I like to do with my life. I like to be outdoors. I like to be out helping people. I just like being out with the guys, doing a job I really love.

"When people say that women shouldn't be firefighters, I say, 'To each his own.' You can listen to someone else's opinion, but it doesn't have to be your reality. Your reality is what you want to make of life. The opinion that should matter the most is yours. If somebody says they don't think women should be doing this, that's fine. Remember, that's how *they* feel. That's not the way *I* feel. It's okay to not agree.

"I hope this book inspires young girls to believe that they can do whatever they put their minds to. Whether it is to be a firefighter, police officer, paramedic, or anything else, I want these stories to give them hope. If reading this book gives you some insight into what you might want to do in life, then come see one of us. Or see someone who works in that capacity in your town and talk to her about how to take it further. Follow up on your idea for yourself and just go for it."

F.D. N.Y.
009
AMBULANCE

LIEUTENANT KATHLEEN GONCZI

AGE 40
EMS LIEUTENANT/PARAMEDIC
FIRE DEPARTMENT OF NEW YORK
EMERGENCY MEDICAL SERVICES COMMAND
BATTALION 16
HARLEM

Even though she wasn't scheduled to work that day, EMS Lieutenant Kathleen Gonczi was among the thousands of rescue workers who rushed to the scene of the World Trade Center disaster on September 11. As a paramedic and EMS supervisor with 14 years on the job, she knew how to organize patient care in a Mass Casualty Incident and felt a strong calling to contribute her knowledge and experience to the rescue efforts at Ground Zero.

One of the most difficult challenges Kathleen faced was keeping the people under her command safe in an unstable, dangerous, and constantly changing environment.

"People who work in EMS are lifesavers," she told us, "but we don't generally risk our lives. That has never been our job. But the ballgame changed on September 11. People were so caught up in wanting to get out there and find someone alive in the rubble that sometimes they just weren't thinking about their own safety. Keeping them safe was the most important aspect of my job."

Although the events of September 11 still haunt her, Kathleen is grateful for one thing: "I was fortunate that I didn't see any human carnage," she said. "I know a lot of people who are going through a terrible time because of what they've seen."

"It was pretty disheartening that there were no patients. It was all ash and metal. There were no people. There were no bodies. There was nothing. All of a sudden I realized that I was breathing in their remains. It was a horrifying realization."

—Lieutenant Kathleen Gonczi, November 5, 2001

"I've been on the job for 14 years, and I worked most of my career on an ambulance during the height of the AIDS and the crack epidemics. I worked the Upper East Side and Upper West Side my first four years, then I went to Harlem. It was not unusual to respond to five shootings on a Friday night when I first was an EMT. I've been a boss now for six years, and every day I go out in a command car.

"Until now, I really thought I'd seen everything. I've seen suicidal jumpers from tall buildings, people under trains who were quite mutilated, even one woman who literally laid herself across the tracks and just wanted to end it. The old-timers fall into autopilot on calls like that, but the rookies don't have that experience yet. That was a very big concern on September 11, because this was their first exposure to anything on a grand scale. It was the worst thing anyone had ever seen.

"I live out in Pennsylvania, and on the morning of September 11, I was awakened by a phone call from a friend who said, 'I think this concerns you. A plane just crashed into the World Trade Center.' I had just finished working three 16-hour shifts. I was supposed to have gone to court with a friend that morning, but I was so tired, I called him the night before and said I just couldn't go. If I had, I would have been in Manhattan that morning, and there's no question that I would have been down there. Everybody who was within earshot was down there.

"I had a lot of reservations about going into Manhattan, but I also felt a call to duty. You don't even know you have it until it's there. I love this job. No one loves this job more than I do. I can't believe that I'm so lucky to have a job where I can't wait to get to work. I have always felt that way. But this day, I was angry that I had this job. I had chest pains, and I got sick to my stomach.

"But I knew I had to go. I have two daughters, 17 and 19, and I've been a single mom for a long time. They grew up with my job, and they know there is an element of danger to it. I called my oldest daughter, Tara, at Penn State, just to make sure she was okay. She said, 'I'm fine. It's very quiet here.'

"Then the plane went down in Pennsylvania, and I didn't know where we were going nationally with this. My youngest daughter, Kristin, is a senior in high school, and I didn't want my daughters separated if something was going to happen to me. I called my oldest daughter again to ask her to come home, and she said, 'I'm on my way. They're closing the school.' She stopped and picked up Kristin, and they were home within a few hours. I couldn't have gone until I saw them.

"I'm usually not one to run toward danger. I always look at the scene before I run into it. I'm the one who's always saying, 'No, we're not going into that burning building until it's secured.' I've only had one experience where I risked my own life without thinking, when I jumped onto the tracks to save a four-year-old who had fallen. By the time I went down there and got back up, I realized that the train had come in.

"I had sworn to my family, to my girls, to my sisters, and to my brother, that I wouldn't go down to the site. I said I would stay at the station. But the closer I got, the more I knew I had to go down there. I knew I could handle it. I knew I wouldn't lose control and that I would be an asset. I had the training, I had the certification, and I had the equipment and resources. Not to mention the fact that I'm from New York, and I've worked in the city my entire career. I have so much more knowledge than people from out of state or the new personnel. I just felt that I was needed down there, and I was going into it with a very clear head.

"That night, I ended up taking command of the triage and treatment area inside the Liberty Plaza Hotel lobby, which was overlooking the site. I was in such amazement and awe. I was stunned more than anything. I remember being concerned about who was down there. I had been told that my chief had passed away, and I was shocked to see him right off the bat. I had people all the way to Friday coming up to me saying, 'Oh my God, you're here!' A lot of people thought I was dead because the sound of my last name is very similar to Fire Chief Ganci, who died in the first collapse. There were a lot of rumors. The closer you got to the site, the less you knew about what was going on. It was very hard to keep track of anybody.

"The Liberty Plaza Hotel was connected to the Brooks Brothers clothing store, and the entire place was destroyed. You could not set up a movie scene to make it look more devastating. It looked like I'd entered a horror movie after a nuclear attack. Nothing looked new. Everything looked very old and worn. I could tell it had been a very upscale place, and the lobby of the hotel was probably fully marbled with a front desk. But now everything was covered with ash, inside and out. There was medical equipment thrown all over the place. They had some food and an eyewash station set up in the back, and there were three gurneys right when you walked in with full medical teams stationed at each one.

"I relieved another lieutenant who had been in command since the collapse, and he looked beat. He handed me a napkin with a list of the ambulance units that were there and said, 'This is what you have.' I probably had eight ambulances ready for transport but only one FDNY ambulance. The rest were from Jersey and other places.

"I was trying to pull everything together. The command wasn't really set up as of yet, understandably, so we were just trying to make sure we had people where they needed to be. The worst realization was that there were no patients. The medical teams got very frustrated because they were sitting there with nothing to do. People got very caught up in wanting to help. They wanted to be out in the rubble, and that's no place to treat patients. My goal was to keep everybody safe. We needed the rescuers to bring people out of the rubble so that we could treat and transport them.

"For 19 hours, I was on call every second. There were a lot of rumors, but most of them turned out not to be true. 'We found people, they're trapped, we can hear them, and any minute we're going to be transporting five people.' We were always ready. During that entire time, I probably sat down for a total of 15 minutes. I had to be up. I had to be on top of it. I never slept. I had one granola bar the entire time. I couldn't eat. I had maybe one bottle of water. I never took my helmet or my turnout coat off, because I felt they kept me visible. Three days later, I could still feel the sensation of my helmet on my head.

"I was awake all night. I was floored, and more than impressed, by the ironworkers and their dedication. One of them actually lacerated his cornea and wanted to go back and work. It took a lot of talking to get him to go to the hospital. I think we treated one firefighter who was injured, but the rest were just bruises, scrapes, and a lot of people needing eyewashes.

"It was pretty disheartening that there were no patients. It was all ash and metal. There were no people. There were no bodies. There was nothing. All of a sudden I realized that I was breathing in their remains. It was a horrifying realization.

"At a quarter to 9 the next morning, I was very aware that we were at the 24-hour mark. A lot of people had fallen asleep. We had no blankets, so they handed out all the coats from Brooks Brothers for people to sleep on. I was the only one awake at that hour, and over the Citywide frequency, I heard a dispatcher declare a Mass Casualty Incident at Lincoln Hospital because the generator blew up. I panicked, because I thought things were starting all over again, and that we were under attack.

"I was relieved of my command at around 5 o'clock on Wednesday night. Ash was very thickly caked on my boots, and I stopped by the station and hosed them down. There's still ash in the seams of my boots, even though I've polished them since then. I find it very comforting that it's there now. It should stay there. I can't explain exactly why, but I know I'm extremely comforted by my boots.

"I went to my parents' house in Queens, and by the time I got there, I was terribly, terribly shell-shocked. The fighter jets came over, and I almost fell off my chair at the kitchen table. I ran outside because I had no idea what that thunderous noise was. I began interrogating little children who were playing outside about what they'd seen. I must have looked like a nut.

"I finally fell asleep around 10 o'clock that night. I set the alarm for 3, and I was back down at the same command by 5:30 Thursday morning. The triage and treatment area had been moved out of the Liberty Plaza Hotel because there was so much debris inside and it was no longer conducive to treatment. Everything had been moved to the outside of the Millennium Hotel.

"The Millennium is a full glass structure, and there were a lot of heavy glass shards hanging off it about 20 stories up. It didn't look safe to me. I found some police tape and cordoned off the front, and a half an hour later, I saw someone eating lunch inside that area. I said, 'This is cordoned off. Look at the glass up there.' I came back another half hour later, and more people were sitting there.

"So we moved to the side of the building, and at that point, I felt that I really wanted to get more organized. I would not take no for an answer, and I was on everybody to get in line. Corrections was driving by with a truck, and they gave us a tent, so I set up a treatment area with doctors and nurses. They were very glad that we were getting it more organized.

"We had probably nine ambulances at that time, most of them from out of state. We had four stretchers ready to go into the site, and each one had a doctor, a nurse, and an EMS team. At one point I turned around, and our only access out was blocked by a tow truck bigger than a house. Somebody had just left it there, and it took us an hour to find out who the driver was.

"There was a system set up where three pulls of a blow horn meant that something was collapsing. It meant drop everything and run. Right after we got a call for the first patient, and a stretcher went in for an injured rescue worker, the siren went off and we saw a couple hundred people running toward us. Some of the people under my command hadn't caught on, and I had to turn around and say, 'C'mon, c'mon, c'mon!' Then I ran. I ran very far.

"I had a safety meeting, and I told them, 'When that horn blows, I'm running. I'm not looking back. And you have to do the same thing.' I wanted everyone to recognize the dire need to run. I said, 'Don't worry about where anything is. Leave everything behind.' I wanted them to be responsible for themselves at that point and to not care about anything else but running.

"I had heard by then that both the Millennium and the Liberty Plaza hotels were very unstable. We were basically between them. I was really concerned about the stability of the area, and I decided that we weren't in a safe place. It was very nerve-wracking to go back in and get our equipment. I was responsible for the safety of a lot of people, and the last thing I wanted was to have somebody else get injured.

"We set up again, and again the sirens blew and everybody started to run. We kept ending up farther and farther away, and then we'd come back in. Degree by degree, we were being pushed out of there because of safety considerations. I had to draw the line between how much the operation was worth, and how much personal safety was worth. At some point, there was a bomb scare one block over from where we were. It seemed like every time we felt that we'd gotten into a safe position, there was another catastrophe.

"When I went for my lieutenant's training, we had a little city set up inside the Academy. It was called Sternville, named after one of the officers who designed it. They put a helmet on you and gave you a radio, and anything that could happen did happen in Sternville—high-rise fires, helicopter crashes, river rescues—all at the same time. This event definitely was Sternville. We constantly had to deal with one problem after the next.

"I think I was able to implement my training in terms of organizing the different sectors. A guy on a bicycle came up to me and said, 'What can I do?' I put him in charge of registration. Just take everybody's name. There was a line of people, and we wanted to keep a count of everybody. I don't think it was a perfect operation. The communication was the problem more than anything. We only had one radio, so we didn't have any way of contacting each other. Any time I sent an ambulance down to the site to pick up a patient on a stretcher, we never heard from them again.

"I got relieved about midnight, and by then they'd put up an iron curtain around the Millennium Hotel to prevent the glass from falling. I felt justified in my complaints that it was not safe to operate in that area. I couldn't have lived with myself if people had gotten hurt when I was responsible for them.

"I was pretty beat up. I couldn't walk for about four days because I'd run so much. I got very sick with severe respiratory infections, and I had to go on antibiotics. For the first six weeks, I was having very worried sleep. This has not changed me for the good. I feel older. I feel like I've aged a lot. I've always used humor in just about every aspect of my life, and I don't do that anymore. I think things are a lot more serious. I feel a lot more responsibility toward my daughters.

"My first day back on patrol, my first call was a bomb scare at the George Washington Bridge, which I'd just driven over to get to work. It turned out to be just a backpack that someone had dropped, but it was very stressful.

"I've never thought about quitting this job before, but the first few weeks after September 11, I did consider it. I come into work now because I know I have to. It's the nature of my job that any minute all hell could break loose, but this was very different. Each day I go to work now, I'm not sure if I'm going to come back home.

"What they did to us was beyond humanity. I think they took advantage of our system, and we were caught from behind. They got the best of us just based on the fact that we're better natured in life. I have a lot of anger. It's anger that's beyond my scope of tolerance. I felt the anger from all the cops and firefighters down there, too. They were so mad. They were so angry. It's not just that 3,000 people perished. It's that 3,000 people were mutilated, and they will never have a proper burial.

"A lot of the EMTs and paramedics I supervise want to debrief with me. One woman lost three members of her family— innocent people who were just going to work were killed to prove a point. There are a lot of casualties of war, and I think you expect that when you enter war. But we weren't at war. I don't think we expected anything like this."

20717
MANHATTAN
20717
RAMOS

NANCY RAMOS-WILLIAMS

AGE 36
POLICE OFFICER
NEW YORK CITY POLICE DEPARTMENT
MANHATTAN TRAFFIC TASK FORCE

Rookie Police Officer Nancy Ramos-Williams knew she was putting herself in grave danger when she was among the first to arrive on the scene of a high-rise fire in Lower Manhattan on September 11. Nonetheless, she went to work immediately, evacuating people from the burning North Tower as debris crashed down around her.

When the first Tower fell, Nancy thought quickly and logically, seeking shelter against the supporting pillars of Building Five. Even though she escaped immediate death from the fire and falling structure, her life was in peril as she choked on the swirling debris. Had it not been for the quick actions of a stranger, Nancy fears that she might not have made it out alive.

One of the things that preyed on her mind after that day was that she didn't know the name of the man who had saved her. But Nancy was determined to find her rescuer. Several months later, after many inquiries, she was reunited with Court Officer Edwin Kennedy during a surprise meeting on the front steps of the court building.

"I was standing there looking at him," she told us, "and he kept getting closer and closer to me, and then he said, 'Your eyes. There's something about your eyes. Oh my God! It's you! How do you feel?'"

Even though she helped evacuate hundreds of people and stopped to aid others after she was injured herself, Nancy humbly told us that she believes it's Officer Kennedy who's the real hero.

"I said to myself, *Oh my God, I'm a rookie. I'm new. I don't have a lot of years on the job, and I'm still learning.* But then I realized, this is the job that I chose to do, and I wanted to stay there and continue helping people."

—Nancy Ramos-Williams, January 26, 2002

"I've been a police officer for almost two years, and I'm assigned to the NYPD's Truck Enforcement Unit. We work Lower Manhattan and oversee anything that has to do with truck traffic. Our job is to make sure all trucks comply with city regulations. Before I took the test for police officer and went into the Academy, I worked for the NYPD as a civilian traffic agent for 10 years.

"On September 11, I was at police headquarters, maybe 8 or 10 blocks away from the World Trade Center. I'd gone upstairs to do some paperwork, and I was heading to my post with my partners, Richie Vitale and Charles Rubenstrunk. We were waiting for a light when we heard a big explosion. We got out of the van, and we looked around to see what was going on. We saw this huge hole in one of the Twin Towers, and we saw fire. We saw paper coming down, like when we have parades for the Mets.

"So we turned the van around and went through the security system at One Police Plaza, and as we were waiting for the bars to go up, I remember that somebody knocked on my window. I saw this guy trying to show me a badge because he wanted to go down there. I opened the door and said, 'Just get in and let's go.'

"We got there in less than a minute. We parked our van in front of the Millennium Hotel and started running toward Building Five. Everybody from the Tower was coming out through that building.

"There were only a few of us there then. Right away I saw two other female officers, Tracy Donahoo and Carol Paukner, so we worked together. We decided to stay near the entrance and help evacuate people. We were telling them, 'Run! Get out of here! Don't look back!' We were telling them to go straight to Broadway, and we told them not to use their cell phones, because they could activate a bomb.

"My partners were taking in the people who were hurt or who couldn't walk and helping them to ambulances. I had to stop a few times and help people, because the guys were tied up with so many injured people. I remember helping an Asian woman whose hair was burned to her scalp. It looked like a piece of rug. I grabbed her because she was confused, and I took her to the ambulance.

"More people were coming out of the building, and more police officers were coming toward the building. I saw people with no shoes running out of the building. I saw this man walking like a zombie, with no shoes, burned from head to toe. He was very badly burned. There was a lot of blood.

"I said to myself, *Oh my God, I'm a rookie. I'm new. I don't have a lot of years on the job, and I'm still learning.* But then I realized, this is the job that I chose do, and I wanted to stay there and continue helping people.

"I remember seeing a man who was blind, and he had a dog. I wanted to help him, but he told me, 'No officer, I'm fine. I can find my way out.' A lot of people were kind of nasty toward us because we were telling them to run and not to use their cell phones. They were like, 'You don't tell me what to do.' I thought, *You don't know what's going on. You're cursing me out, but you don't know what is up there.*

"I heard on the radio that there was another plane missing, and they had a feeling it was coming toward the World Trade Center. I ignored that, because I was saying to myself, *These people need me, and I'm just going to stay here.* A couple seconds later, I hear a big *Boom!* and felt a shaking. One of the other officers ran toward the street, and I grabbed on and said, 'Don't run to the street! You're going to get killed!' The civilians were running toward the street, and a lot of people were getting killed.

"Parts of the building and parts of the plane were coming down, and they were smashing the fire trucks and the ambulances. We ran around to the revolving door of Borders Bookstore, and we could see the orange flames coming at us. It looked like *Star Wars,* because the orange flames were coming so close.

"The bookstore was closed, and we were trying to kick the revolving door so that we could get in. We could see all these parts from the building coming down, killing people, and they were falling very close to us. I looked to my left, and I saw the Post Office. I saw this huge chunk of rock hit the wall. There was a delivery guy on a bicycle, and that rock, after it hit that wall, came down in the middle of the street and smashed this guy.

"Everything went back to normal, and we could see papers coming down. We were like, 'Okay, the first one hit and just paper was coming down, and now the second one hit and papers are coming down again, so it must be safe for us to go back.' So we went back to the entrance of Building Five, and we were telling people, 'Run! Run!' You could see the fear in people's eyes, and we were seeing more people bleeding and burned.

"I felt the ground shaking like an earthquake. Our bodies were going sideways, and I could hear in the background a *Zoom!* like a plane was taking off. I knew that the base of a structure is always support for the whole building, so I ran into the corner of the building to the first column and I just stood right there. People were running into the street, and they were being killed by falling debris. There was a firefighter in front of me, and he was shaking.

"I could feel people trying to squeeze me in the back and on the side. I had my weapon on the right side, and I put my hand on my weapon because I didn't want anybody to take it and hurt somebody. I could feel the firefighter covering my head with his helmet. Then he kind of cuddled me, and whoever was behind me did the same thing.

"My eyes were closed, and that stuff was coming down, and I could feel the air at the same time going *Woooo woooo* and trying to suck me away from the column. I told God, 'Just take me if you're going to take me.' The air was so strong, trying to suck me out of the area, and I was pushing myself to the wall. I said, 'I'm all yours, God, take me,' because I didn't know what was going on.

"Then everything got quiet and there was silence. You could hear a pin drop. I opened my eyes, and it was pitch black. I said to myself, *Oh my God, I lost my sight! I can't see! I can't hear anything! I can't see anything! I can't breathe!* I started saying, 'I can't breathe!' I felt somebody holding my hand, and he was telling me, 'I've got you. Don't worry, I won't let you down.' Then he said, 'Remember training. Remember training.'

"What that means is you never give up. Our training teaches us that when you're in a situation where you're feeling helpless, you give 120 percent. It's your life. You can't just give up. You have to continue. That's how I knew he was a police officer. He could have just left me there, and I would have died, because I couldn't breathe.

"I was opening my mouth, and the more I was spitting out, the more was coming in. It was choking me, and I couldn't breathe. He told me later that I had been knocked out, and he had given me oxygen. He had a medical bag and he gave me oxygen, and that's when I came back.

"So he pulled me up, and we were walking along the building near the bookstore, between the entrance and the revolving door. He was holding my hand as we walked. Then I stepped on a body. I fell onto a headless body, and my hand just let go of his. He looked for me in the debris and held my hands and tried to get me up, but I slipped again. I had hurt my leg, and I couldn't get up.

"He told me later that my hand was on top of the person's neck, because it was a headless body. All of the people who stood in that area were killed. They were on top of each other with the debris, and we couldn't walk through there, because when the Tower collapsed, it fell on top of Building Five. We were trapped in the middle.

"We made a right turn, and we were facing the back of the church and the cemetery. There was a small gap, and we dragged everybody who was alive out of there through that little hole.

"I remember walking, and now I could see, and everything was gray. I could hear somebody saying, 'Oh my God, I'm bleeding.' I saw this person bleeding from his neck. I saw a telephone on the ground. I saw hands, body parts, and handbags. I saw all that, and I was like, 'Oh my God! This is for real!'

"I was choking and choking, and I couldn't breathe. He was shaking my shoulders and saying, 'Don't give up. Don't give up.' He told me that at this point I stopped breathing and that he had to put his fingers down my throat so I could vomit all that stuff out.

"I could hear the firefighters breaking into some glass, and they dragged me into a deli. Once I was in the deli, I saw people coughing. I was choking, and the officer who was with me was cleaning my face with water. I knew there were a lot of people around me, and I heard somebody say, 'EMS, the officer, her hands are bleeding.' I checked my hands. They were covered with a lot of blood, but I was not cut. It was the blood from the body I fell on.

"Everybody in the deli was dirty, filthy, so I looked at myself in the mirror and I had to clean my nose because I had all that soot. I was confused, I wanted to cry, but I couldn't cry, because I'm a police officer. There were civilians in there, and I couldn't show the civilians that I was shaking. Where were my partners? Whatever happened, I lost them.

"There was a lady crying, and I just had to tell her, 'You know, it's okay. We're here. Whatever happened, we're alive. You're going to go home to your family. You're going to be okay.' I looked toward the door, and I saw so many body parts around the area. As I was getting ready to go out, a firefighter stopped me and said, 'Officer, there's nothing for us to do out there. We did our best.'

"But I was wondering, *Where are my partners? Where's Carol? Where's Tracy?* I still didn't know that the building had collapsed. My chief happened to run into that deli with two lieutenants. He asked me, 'Are you okay?' I said, 'I'm injured and I can't walk. I don't know if my leg is broken, but I can drag myself.' He said, 'I want you to drag yourself outside, make a left, go to Broadway, and get out of here. You're going to see some cars out there, and you're going to sit down in one of them.'

"So that's what I did. I dragged myself all the way up to Broadway, just dragging one leg. As I was making a left on Broadway, I saw an EMS worker coming out of a building, and she was stumbling from one side to the other. She said, 'Officer, help me! I'm dying!' She could hardly talk. So I grabbed her, and she collapsed in my arms, unconscious. She was very petite and not very heavy, so I held her with my left arm. Then I flagged down one of the ambulances and asked them to take care of her.

"I kept walking until I saw one of our Parking Enforcement trucks. My old supervisor from Traffic was inside. I sat down in the truck, and then I saw my chief running toward me. He opened the door, and there was a big mushroom cloud after him. He jumped in the truck, we closed the door, and we were all coughing because a lot of dust came in. It got pitch black. I asked, 'What's going on?' and that's when he told me that the second building had collapsed. It got pitch black outside, and we were still there in the truck, in shock.

"After it started getting clear again, my chief said, 'I'm going to go. Can I use your radio?' I gave him my radio and grabbed a helmet I'd found in the deli and said, 'Chief, you might need this.' He put on the helmet and left.

"The next thing I knew, one of the police officers who works with me came and told me, 'I'm here to take you to the hospital.' Beakman Hospital is probably five blocks away, and he wanted to carry me, but I said, 'No, I can help myself.' I dragged myself all the way down there.

"When I got to the hospital, they hosed me down with water because I was covered in soot. I was in shock. I know the doctor was asking me a lot of questions. Because I wasn't burned or anything serious, I said, 'You can just put something on my leg, give me crutches so that I can walk better, and I'll go to the doctor another day. I know there are people hurt more seriously than I am.'

"We begged one of the ambulances to take us close to our precinct, and we walked the rest of the way. I was on crutches, I didn't have shoes, and my hair was like a pancake because of the soot and the water. I was a total mess. When I walked into the house, an officer looked at me and said, 'Oh my God! Are you okay?'

"My tears started coming down, and I said, 'There are so many body parts. So many people died down there, and I couldn't do anything for them. I don't know what happened to Richie or Charles or Carol or Tracy.' All this was going through my head, and it was all coming out.

"I took a shower, and someone called me and told me that my partners were alive. I went downstairs to the basement, and I saw my partners and Carol, and she was crying. Carol's partner wasn't there, but we found out later that she was safe, too.

"When the whole thing was over, Richie told me that when the building collapsed, he said, 'Nancy, hold on to my arms.' He was thinking that the person he was helping out was me. He said it was pitch black, and when he got outside, he looked back and saw that he didn't have me. He had Carol. He said, 'At least I was happy that I had another police officer with me and that we saved each other.'

"We lost a traffic officer from our division. The guy I picked up, the one who knocked on my window at the beginning, never made it out. He was a volunteer for the ambulance. So inside of me, I still feel guilty that he never made it out. When I'm at my family reunion and there's happiness and joy, I feel sorry that he's not there with his family to do the same thing. That really hit me. I have that right here in my heart.

"I'm a Christian woman, and now I appreciate life even more. I've got two kids. Keith is 17, and Nastazia is 10. I still love my job. I just want my life to continue to be happy.

"September 11 changed my life a lot. It changed it so much that I got married. Christopher and I were supposed to get married in June, but we kept postponing it. Then with September 11, it got worse, because he's a civilian for the police department and we were both working at Ground Zero. He had crazy hours and I had crazy hours. Finally we said, 'Let's do it.' And we did."

LIEUTENANT

CAPTAIN BRENDA BERKMAN

AGE 50
CAPTAIN
FIRE DEPARTMENT OF NEW YORK
LADDER 12
CHELSEA

In the weeks and months following the terrorist attack on the World Trade Center, Brenda Berkman was among many of us who noticed a lack of media coverage about the women who responded to the tragedy.

By the time we caught up with Brenda four months after the incident, she had launched a widespread campaign to increase the visibility of women at Ground Zero and women emergency workers in general. She'd been to Washington, D.C., where she'd addressed some of the country's most prominent feminists and politicians at a dinner sponsored by the National Women's Law Center. She'd met with congress-women and judges. She'd generated national media attention. And she'd worked with the National Organization for Women's Legal Defense and Education Fund to make a 12-minute video, The Women of Ground Zero, *which is still being shown around the country.*

It was difficult to track Brenda down, to slow her down, and to sit her down for this interview. As one of only two regular officers at a firehouse that lost five members on September 11, she has been consumed with funerals and memorial services, caring for families, training new personnel, and responding to emergency calls. Eight months after the attack on the World Trade Center, she was promoted from lieutenant to captain.

There's no question that Brenda is tough. Her 20 years in the fire service and trailblazing ways are proof enough of that. But there is a soft side to Brenda, too, as was evidenced by the tears that fell for her fallen comrades when she spoke about them during the interview that follows.

"I was in command of this group, and I had nobody to command me, and all I could think was, *This is what Vietnam must have been like.* You're just out there in your little groups, and you're left to fend for yourself."

—Captain Brenda Berkman, January 24, 2002

"September 11 was my day off. It was Election Day, and I was going to go out and do some campaign work for a candidate I was supporting. I was still at home in Brooklyn when I got a phone call saying, 'Turn on your television!' There was a replay of the first plane going into the Tower, and I immediately started putting on whatever parts of my uniform I had at home. My first thought was that I would go into headquarters. This was going to be a big high-rise fire. They were going to need a tremendous amount of resources, and I thought, *There's got to be something they're going to need that I can do for them.*

"I knew that it would be a problem getting to my own firehouse in Manhattan, and when I got out into the street, they were shutting down all the buses and subways in Brooklyn. I decided to go to my old Brooklyn firehouse instead, and by the time I got there, they'd issued a general recall of all firefighters.

"So I was at this firehouse with a bunch of people I knew from my old company and other people who had also come in to work, and we started getting ourselves together, getting gear, and getting whatever tools were left over. I borrowed the captain's spare set of gear because he was about my size. There really wasn't much else left, because everything had gone out with the trucks. Every company from that section of Brooklyn had responded already. That's why they suffered such tremendous losses.

"We commandeered a police van and headed for the convening point where they were telling all off-duty firefighters to go. This cop was driving us with lights and sirens and everything going. We passed another firehouse where more people were leaving to go, and we led the charge into Manhattan with this whole group of Brooklyn firefighters behind us.

"We were going across the Brooklyn Bridge, and our side of the bridge into Manhattan was completely empty. The other side of the bridge was wall-to-wall people walking. A lot of them were covered in dust. When we got down to the bottom of the bridge on the Manhattan side, it was a dust storm. This was probably half a mile from Ground Zero, and the cop had to turn on his windshield wipers to see. We were inching along, inching along. By then, the Twin Towers had fallen, and papers, dust, and other things were starting to pile up. There were vehicles all around. I heard car alarms. I heard explosions. But I heard no people. Somebody said, 'Let us off right here,' and we hopped out.

"I went over to report in to a temporary Command Center. There were a bunch of chiefs and aides there, some of whom were covered in dust. They were ordering, 'All company officers over here. All company firefighters over there.' So I went over to where they were talking to the officers, and they said, 'Do you have anybody from your firehouse here?' I said, 'I don't know.' They said, 'Well, go find them and then come back.'

"There was a long line of firefighters there waiting for assignments. I found quite a lot of my guys who had come down from the firehouse, and they had a few hand tools with them. I found a guy who was on light duty from my firehouse who was there in a pair of shorts and a little blue plastic helmet. He also had a radio with him, so I said, 'You're coming with us, and I'm taking your radio.' I split up from the guys from my old firehouse in Brooklyn, took command of the guys from Chelsea, and eventually we got an assignment and headed out.

"We couldn't even read the street signs. The streets were all torn up, and there were burned-out vehicles all around. We were supposed to go over to the east side of the World Trade Center, on the Church Street side. They said there was a building on fire there. Well, there were many buildings on fire there. Firefighters were stretching lines and trying to operate a tower ladder onto a building that had fire on every floor, out of every window. There was no water pressure. There were no working pumpers. The hydrants were all knocked out.

"We no sooner got going on something there when a chief came along and said, 'Everybody's got to leave the area. We're afraid that Seven World Trade is going to fall down.' The whole south side of Seven World Trade had been hit by the collapse of the second Tower, and there was fire on every floor. Big parts of the building had been demolished and taken out.

"My main concern was not losing the guys who were with me. That was a huge problem. The guys wanted to run around and do stuff, but I couldn't send them out because we had no tools, no breathing apparatus, and no radios to communicate with each other. I was mainly concerned that they'd go off somewhere and get hit by a falling building or that they'd disappear into the rubble and never be heard from again. I was in command of this group, and I had nobody to command me, and all I could think was, *This is what Vietnam must have been like.* You're just out there in your little groups, and you're left to fend for yourself. There was no structure. There was no organization. There was no equipment.

"We started talking to people and asking, 'Where's 12 Truck?' We ran into a guy from Engine 26 who was there at the time of one of the collapses, and he said he had seen 12 Truck over on West Street. So since we'd been told to leave the collapse zone, I made a command decision that we were going to walk over to where 12 Truck had last been seen. We wanted to get to our rig and get whatever equipment was left on it, because there were no other usable fire trucks around us. Everything was stripped or burned or buried. I said, 'Let's get over there, find our guys, get some radios, and regroup.'

"Then we started hearing about companies that were missing. With the extent of the collapse area, all I could think was that there had to be thousands of emergency people and civilians trapped in there. I couldn't see for the smoke. I couldn't see for the piles of rubble. I couldn't get close enough, because of the fires, to really get an idea of where the voids were or where we should start looking for people.

"We got over to West Street, and it was just as chaotic. We tried to find a Command Post, tried to find somebody who could give us an assignment or direct us to where 12 Truck had last been seen. I saw a former battalion chief, and all he could tell me was that Orio Palmer, a battalion chief from our battalion, Battalion 7, was gone. Missing. I cared a lot about Orio. He was a super guy. I thought he was going to be chief of the department someday. He was just the best.

"I kept seeing other women firefighters practically from the moment I got there. I saw Ella McNair over on Broadway, and we gave each other a hug as if to say, 'Thank God, you're alive.' I saw one of our most junior women firefighters working with the tower ladder over on Church Street, then I saw another one of our very junior firefighters. I just kept running into them and clicking them off in my mind. I knew that Rocky Jones was on vacation, but I knew there was potential for a lot of the women to be there. So many of us work in Lower Manhattan and Lower Brooklyn, which were the companies that lost the most firefighters.

"Like I said, West Street was in chaos. That's when we discovered that 12 Truck was missing, and Engine 3, which is also at our firehouse, was missing. All the companies from my battalion were missing, all those firehouses in my battalion. A battalion is a group of companies that cover a certain area, so those are pretty bad odds. I kept seeing guys I knew who said, 'My brother's missing,' or 'So and so's missing.' It was unbelievable.

"But at that point, there was still hope. We thought, *All we have to do is find them. They're in the voids. They gotta be in there somewhere.* We knew right off the bat that Father Mike Judge, the fire department chaplain, had been killed, because somebody told us that. When we were over on Broadway, we ran very quickly into the church where they had brought his body, we paid our respects, and ran back out again.

"Then we ran into a couple guys from my firehouse, from Engine 3 and 12 Truck, who were alive, and they told us the names of the people who were missing. Nobody knew where anybody was. There was so much confusion about where people were assigned and where they ended up. People were blown across the street, or they escaped the first collapse but were caught in the second. My company didn't even realize for days that they were in the Marriott Hotel when the North Tower came down. They thought they were in the South Tower, which had already fallen.

"I found out later that eight officers from my battalion responded on duty. Five of them were killed. An off-duty officer from my battalion was killed. A captain of one of the truck companies from my battalion was killed in the first collapse. There was a huge percentage of people lost from the 1st, 3rd, and 11th Divisions, all of which were the divisions I worked in as a firefighter and officer. I worked in those three divisions that were the hardest hit.

"I had this phenomenon where I'd be looking at the pictures in the newspaper, one right after another, and I'd sit there and try and figure out whether I knew this guy or not, because a lot of times I don't know a guy's name but I just know who he is, right? I worked with him, maybe had a laugh with him, maybe had a meal with him, maybe heard a story about his family, he might have driven me as a chauffeur, or whatever. Some of these guys I felt very close to, and some of them I just had worked with. I'd be looking along the line of pictures and my mind would play games with me. I'd think I was looking at every picture, but I was actually skipping, because I'd be focused on the guy I recognized five pictures back.

"It took me weeks, maybe months, to actually realize that I knew some of these guys. Now I can't even remember who died, there were so many. I keep thinking I'll see them. I'll be sitting somewhere, and some guy's face or his name will flash into my head, and the disbelief just sets in all over again.

"For several weeks, we were all in desperation mode. We had to find those people who were trapped, and that took precedence over everything. Our own physical needs, our families, our outside interests, our second jobs—everything else was put away. Everything I was doing prior to September 11, boom, it just went away. Our sole mission was finding those people and taking care of one another.

"For the women, it was initially a tremendous bonding experience with the men that a lot of us had not really felt before. There was a huge level of grief and loss we were all feeling, both the men and the women. The sheer magnitude of what had happened had a horrible impact on the FDNY. People drew together, and they were very supportive of one another. All over New York, people were sympathetic in ways that I'd never ever seen New Yorkers be toward one another. They were kind, they were considerate, they were loving, and they were giving. I'm not saying that New Yorkers aren't that way, but they don't project that way. They're not touchy-feely. They're brusque, they're in a rush, they don't take time to sit down with somebody and share a quiet moment. But all this was going on, not only within the community, but within the fire department as well. The women and the men were very tight.

"And that was a beautiful thing. Thinking back, I just wish it could have lasted forever. But it couldn't last forever, because all the problems that our organization had, that our city had, that our nation had prior to September 11 still existed. And in some cases, they were made even worse. To see that closeness sort of drift away, that's hard.

"Those of us who survived, those of us who are trying to get a little perspective on this, we're worried about what's going to happen a year from now. What's our department going to be like 5 years from now or 10 years from now? What's my personal health going to be like 5 years from now or 10 years from now when we can't seem to get straight answers from anybody about what was blowing in the wind down there? We didn't give a you-know-what about what was blowing in the wind initially, because people put their own personal safety aside and did what they felt had to be done.

"I think company officers, like myself, continue to feel very concerned about the members of our department. We're concerned about their mental state, their physical state, their distraction level, their sorrow, their grief, and how they're handling it. We're concerned about whether or not the new members are getting the right training and direction, because things are not normal. Now we're hiring a huge number of people, we're promoting a lot of people, and the leadership at the top is changing because so many were killed on September 11. There are a lot of things to be worried about out there.

"At the same time, the women in the fire department have had this other thing layered on, where we've started to feel invisible in the whole picture. Our grief and sorrow and efforts were being totally ignored. Initially, nobody had the time to even think about that kind of thing, and later, we were very worried that by saying anything about our invisibility, people would get the impression that we were looking to be on television ourselves, or some nonsense like that.

"It got to the point where I was going to funeral after funeral of my friends and co-workers, and the eulogists and the cardinal and the mayor and everyone would talk about the men who loved this man, and the men who are grieving, and the men this and the men that, and here I am, sitting in the front. He was my firefighter or my officer. He was my friend. I knew him for 7 years, or 10 years, or 20 years. It's just so wrong.

"Now, admittedly, there's not a huge number of women in the fire department. There are only 25 of us out of 11,500 firefighters. But there's a very high percentage of us who were in houses that lost large numbers of people and who were close to these folks. There were women firefighters coming in from other parts of the country to attend our services and to work as part of the search-and-rescue teams, not to mention all the women cops and women paramedics and EMTs. There were two women cops who died and one woman EMT who died. It was like they never existed.

"What's so distorted, I think, is that the people who are seeking to point out that there were women and people of color at Ground Zero who were doing their patriotic duty the same as white men are the ones who are accused of being divisive. I think it's exactly opposite. I think the message we should be sending out to the world is that our country welcomes people with many different points of view, from many different backgrounds, and with many different abilities, and that we all work together in this country without regard to race or gender. By showing women and people of color as patriots who have contributed to this effort and made sacrifices on behalf of this effort, I think we're only showing ourselves to be stronger than any sexist, racist, or hate-mongering group. This is what we should aspire to show the world, that we have a way of life that deserves to be preserved. I think that's a patriotic message.

"But we have to actually behave that way, and if we don't show the diversity of people who are involved in the effort, people will feel that this is all just a put-on. Americans are complaining about the treatment of Afghan women, but how many really respect the abilities of their own women, right here at home? How many Americans really do accept the Muslims? How many really do incorporate a multicultural society?

"I think that people who are trying to raise America's awareness of the contributions of many different groups are doing a patriotic thing. We've made mistakes like this before in this country, where we've sought to erase people from history. Kids never learned about the rich contributions of Native Americans or women or other groups to this country. They didn't learn for a long time about our very shabby treatment of Japanese Americans in this country in the time of WWII.

"What I don't want to have happen is that the efforts of the women at Ground Zero end up being some grad student's research project 25 years down the line because nobody had ever heard of us or because nobody had ever heard or seen a woman talking about her role in this piece of our history. I get e-mails from people who say to me, 'I've done an Internet search looking for stories about women's contributions at Ground Zero and the fight against terrorism, and I can't find a single article. I can't find a single news story. What's with that?'

"When I was growing up, you didn't see women firefighters. Those of us who have spent 25 years trying to make it clear to the American public that women are firefighters, that we do our jobs well, that we are good, contributing members of the fire service, suddenly find ourselves shoved right back in the closet. Once again, nobody knows that we exist.

"I think it's important for people to know that women were there, because these women were dedicated, they made enormous sacrifices, and they continue to make enormous sacrifices on behalf of the fire service and the people they serve. To overlook those contributions not only would be a disservice to the women in New York and the many women who responded from outside New York, but it would be a disservice to the American people.

"I've had people say to me, 'I am so proud to know that women are contributing in this effort. I have three daughters, and they're excited by the idea that they can be firefighters, that they can do public service for their communities, and that they can contribute to our country.'

"This is something we all should be proud of. These women are living the American dream. They're demonstrating that stereotypes and prejudices and biases and narrowness can be overcome and that people of many different backgrounds can make enormous contributions to our society."

FDNY

CHRISTINE MAZZOLA

AGE 34
PARAMEDIC
FIRE DEPARTMENT OF NEW YORK
EMERGENCY MEDICAL SERVICES COMMAND
BATTALION 43
CONEY ISLAND

Paramedic Christine Mazzola doesn't really consider herself brave. "I was terrified being there," she said of her experiences at Ground Zero on September 11. "Anyone who tells you they weren't scared isn't telling the truth."

Christine may have been terrified, but that didn't stop her from providing medical care to firefighters and other rescuers who emerged from the rubble. One of the things that impressed us most about her story was her ability to listen to herself and act on her own instincts when she felt she'd had enough.

Born and raised on Staten Island, Christine has been honored for saving a number of lives in the field during her 12-year career. She was quick to point out, however, that working on an ambulance is not always as dramatic as television would have us believe. As one of two paramedic technical advisors for the NBC television series Third Watch, *Christine knows the difference between TV drama and the routine of working the streets.*

"A lot of it is just hand-holding and providing reassurance," she said. "It's a good feeling to be able to help people, to touch their lives in a way that others can't. Where else could I get paid to do something I enjoy so much?"

"Not being able to treat anybody because there was nobody left to treat, really took a toll on me."

—Christine Mazzola, November 1, 2001

"I was sitting in my kitchen, drinking a cup of coffee and reading the newspaper, when I got a phone call from a friend who said, 'Turn on the TV! Do you know what's going on?' So I turned on the TV, and I couldn't believe that one of the Towers was in flames. I was in awe, you know, my mouth open, the whole thing. Then I saw the second plane hit, and I knew this was no accident.

"In the midst of all this, I was getting dressed, putting on my uniform, listening with one ear to the TV. I heard that all the bridges and tunnels were closed because this was a terrorist action. I live in Staten Island, so I immediately drove to the EMS station closest to me. When I got to Battalion 22, a bunch of people were grouped together around one little TV set, waiting to be shipped into the city.

"I couldn't have been there more than 10 minutes before we left for Manhattan. I rode with two paramedics, three EMTs, and an officer. We had no equipment, no radios, nothing. We just jumped into the car and off we went.

"As we were crossing the Verrazano Bridge, the second Tower fell. We could see the billowing smoke and everything else coming out. It was just unbelievable, even from that distance. It got worse and worse the closer we got. We came into Lower Manhattan, and there was this stepping up of dust on the ground and on the cars and on everything you could possibly see. It was so desolate. There were no people. Imagine Manhattan with no one around. It was really eerie.

"We parked about eight blocks away because everything was so chaotic. There was no triage system, communications were down, and they didn't have the Command Center set up at that time. They were pushing everything back to about three blocks away from the actual site.

"As we were walking toward the Towers, it seemed as if time had stopped. The donut carts on the street were abandoned, and the donuts were covered in dust. There was a cup of coffee just sitting there, where somebody who was getting ready to pay for it had just left it. People had written things in the dust on the cars: 'World War III has started. God help us.'

"I hadn't even arrived on the scene yet, and this is what I saw on my approach. The only people I did see were emergency personnel, because the civilians had been escorted out or ran out. I saw emergency vehicles and police, fire, and EMS personnel. There was no military on the scene at that time, but the planes were flying over, which was also really eerie. I'd never experienced jet planes flying over Manhattan.

"When we finally got to the MERV, the Mobile Emergency Rescue Vehicle, I wound up getting assigned to that. It has four emergency cots set up and is stocked with medical equipment. We were right next to Stuyvesant High School, where they made the Command Center for all the operations. There weren't a lot of people there yet, but as time went on, everyone started to congregate in that particular area. I'm not sure how much time went by, but at one point I looked up and there were thousands of firefighters, EMS personnel, and police officers, all in the same area.

"Slowly, within the first two hours, emergency personnel started to come out of the rubble. There were other triage areas for civilians all around the site, but we were there as the main triage area for emergency personnel. Little by little they started trickling in, the people who were escaping from the rubble where they'd been trapped. They couldn't breathe, they had all this dust and crap in their eyes and in their nostrils, and they were wheezing and having asthma attacks. There were a lot of serious respiratory problems.

"But for the most part, the emergency personnel who were able to get out were okay. There was a huge extreme: Anyone who escaped was pretty much fine, and anyone who was in there was dead. There was no middle ground.

"After the first fire, there were hundreds of civilian injuries. People were treated for pretty serious burns, amputations, eviscerations, lacerations, and things like that. But after the collapse of both buildings, there was almost nothing to do. That's what really freaked me the most. We were there for such a long period of time, expecting thousands and thousands of injuries, and nothing was coming. Nothing was coming. That's the kind of thing that smacks you in the head.

"The doctors and nurses set up a makeshift hospital, like a M.A.S.H. unit, and that was pretty eerie in itself. There were rows and rows and rows of cots set up. They were all empty. The whole thing was about waiting. That's what ripped most people apart. We were there to help, and there was nothing we could do.

"There were people rescued in the first two days, but they were few and far between. It sounded great when you heard it on the news: 'Another person rescued!' But nobody explained that six hours had elapsed when nobody was coming out. Things like that were taking a toll on everyone.

"We didn't have any contact with the outside world, so I didn't know what was going on elsewhere. I didn't know whether World War III had actually started or not. My family. My friends. Were they safe? I knew that I was okay and that I was safe for the moment. But I had no idea what was going on in the rest of the world until I could get my hands on a cell phone and speak to my family. I knew the Pentagon was hit, I knew the Towers had fallen, but I didn't know what to expect next. A missile?

"There was a large group of firefighters rotating into the rubble to help pull people out, and the faces on those firefighters were just numbing. They were expressionless except for the horror in their eyes. And they were not able to release their emotion. It was locked up, held in, and nobody really wanted to express how they were feeling at that very moment. They needed to be strong, they needed to do what they had to do, they needed to help rescue and pull people out. Nobody was talking.

"There was word going around about how many were lost. People were talking about entire engine companies, ladder companies, and rescue companies that were missing. All these numbers started to add up. 'Did anybody see this one, did anybody see that one?' It was just a whole day of 'Did you see this person?' As I was treating people, they'd be saying, 'I have six people in there. I have eight. I have 16.' These were firefighters who were in serious respiratory distress and needed to go to the hospital. But they refused to leave because they had to go back in and save their friends.

"After being there for maybe eight hours, I finally had a break. I was able to sit down quietly up on the fifth floor of the makeshift hospital inside the school, have a little something to eat, be in my own thoughts, and try to unwind a little bit. With the heat that was coming off my body from being outside, the wind whipping, and everything else, there was this horrific smell coming off me. Have you ever been to a tanning salon? It was 10 times worse than that. It had gotten absorbed into my skin and was coming off me when I warmed up.

"I knew what it was. I knew exactly what it was as I sat there trying to eat a hamburger. Jet fuel burns at over 2,000 degrees, and when the Towers were first engulfed in flames, anyone in even close proximity was incinerated instantly. The elevator shaft acted like a vacuum, drawing everything down. A huge fireball exploded in the lobby, and people were just covered in flames. Those were the initial injuries. All burns. Serious burns. And then afterward, when the collapse started, people were trapped within the rubble of concrete and melted steel and everything just kind of pancaked together. The possibility of humans surviving that …

"But I didn't want to think about that. I didn't want to think that people weren't going to walk out of there. We were lucky enough to get 25,000 people out of the buildings initially. This was at 9 in the morning on a Tuesday. I mean, even if you don't go to work on Monday, you're there Tuesday.

"For those who didn't get out, if they didn't jump, if they weren't burned, if they weren't trapped in the collapse, they were blown apart by the explosion. A very good friend of mine had a torso land right in front of her. From the sky. Arms and limbs gone, head gone, just a torso. This is not the kind of thing people like to talk about. I always wondered why my father never discussed going to Vietnam. He would talk about his buddies, he would talk about his training, but he never discussed actually being there and what went on. After this, I can understand why. It's not the kind of thing people discuss, seeing fingers and toes.

"At one point later in the day, I went down to the site to see it for myself. As I was standing there, what really struck me was the silence. Living in the city all my life, I've never imagined what silence would be like in the middle of Manhattan. I heard back-up alarms on emergency vehicles that were blocks away. The echo of somebody barking an order. People just walking, not saying a word. You could hear their footsteps in the rubble and the dirt. It was nighttime, and the entire site was lit up with lights. It was so eerie with the fog and the smoke. It was just unreal. It was like a dream. I couldn't believe I was actually standing in the middle of this.

"I give the women a lot of credit for getting in there and digging alongside the men. I have a lot of respect for our women firefighters and police officers, not being given the recognition they deserve and just kind of quietly pulling up the rear. A lot of people describe what it was like digging and being on the pile, but I didn't want to be a part of that. I was there to help people, and that's what I was hoping to do. Not being able to treat anybody because there was nobody left to treat really took a toll on me.

"There was a point when I knew that there was no reason for me to be there anymore that night. I was there for more than 16 hours that first day and into the next, and I didn't go back after that.

"Being a paramedic and working on an ambulance, I had the option of not going back. They were asking for volunteers to come in on their days off because they didn't want to take personnel from street. We still had the rest of the city to take care of. I decided for my own mental health that I couldn't go back. After the second day, there really were no whole bodies coming out. I didn't want to see body parts. I didn't want to see things like that.

"What happened on September 11 and going back to work on September 12 were two different things. Two different days. There's a part of me that feels grateful to have patients to treat. It feels good taking care of someone again. I guess getting back to business, if you will, and being able to treat the regular everyday Joes feels like getting back to daily life. We need to get back to daily life. We need some sort of normalcy. Normalcy to me means you getting sick and me picking you up. I feel good doing that.

"I recently lost my mom, and that was an earthshaking experience. My mom and I were very, very close. I didn't want to do this job after that, and I didn't feel like I could go on. I mean, how could I treat somebody else when I was broken myself?

"September 11 put that back into perspective for me. It really turned me around. It made me realize that this is where I'm supposed to be, this is why I'm here. I have to believe that there are reasons why I wasn't injured or trapped or killed on that day. There's got to be more. There's got to be something more, and I've got to trust that it's something important.

"Interestingly enough, I haven't had a problem sleeping. I haven't had dreams, I haven't had visions, and I haven't had the daily horrors like some other people have. Maybe it's because I was strong enough to know when I'd had enough. I was there to do the initial job, and when I knew there was nothing left for me to do, I went home.

"Now I'm back in my ambulance, back on the streets of Coney Island. That's where I feel the most comfortable and the most useful. The first time I put this uniform on 12 years ago, I knew it was meant to be. I was meant to help the living."

POLICASTRO

SERGEANT CAREY POLICASTRO

AGE 33
SERGEANT
NEW YORK CITY POLICE DEPARTMENT
BROOKLYN NORTH

Six weeks after the September 11 attacks, Sergeant Carey Policastro took us to Ground Zero to see the devastation for ourselves.

"Coming down here today is like ripping open a wound," she told us as we walked among the dump trucks and cranes. "It doesn't matter how many times I've been here or how many times I've seen it. Each time I come back, it's like the first time."

Carey was studying to become a physical education teacher when she heard a recruitment ad for the NYPD on the radio 12 years ago and decided to give it a shot. As a cadet, she used her teaching background to run an NYPD summer softball league and winter basketball program for inner-city kids.

Today, Carey remains civic-minded and active in her community as a member of the board of directors of the NYPD's Policewomen's Endowment Association. After the loss of so many colleagues on September 11, she spearheaded a fund-raising project to collect more than $135,000 for the families of firefighters and police officers who died at the World Trade Center.

Carey was generous and kind to us, as well. When we ended up in New York over Thanksgiving to conduct interviews for this book, she invited us to her mother's house in Queens to share a holiday dinner with her family.

"There are 3,000 people lying in a pit. There are 3,000 families, hundreds of thousands of people who knew, or loved, or lived with these people. It's the biggest grave this country has ever seen."

—Sergeant Carey Policastro, October 30, 2001

"I work for the two-star chief in charge of Brooklyn North, which encompasses 10 commands, 10 different precincts, Housing, and Transit. I'm his right hand. I handle any newsworthy events, police-involved shootings, community concerns, or whatever needs to be done on any given day. If he goes out to one major event, I'll go out to another major event and report back to him and keep him abreast of what's going on.

"This is the first time in 12 years on the job that I've had an inside position. I came off the street about a year ago to complete my Master's degree. I started out as a patrol cop in the 110th Precinct in Queens, worked a plainclothes anti-crime unit, and then went into Narcotics. I was an undercover cop uptown in Harlem for three years, purchasing drugs. After I got promoted to sergeant, I worked in several Queens precincts, and then went back to Narcotics to supervise buy-and-bust operations. I ended up running operations for this chief in the Detective Bureau, and when he got promoted, he took me along to help him out in his new position. When a two-star asks you to go, you don't say no.

"On the morning of September 11, I was on a plane coming back from West Palm Beach, Florida. I'd made a weekend hop down there, took Monday off, and was planning to go from the airport straight to work. We were only about 25 minutes out of New York City when the pilot came on the air and said, 'There's some terrorist activity in New York City, so we're going to turn around and land you guys in Atlanta.'

"I got on the air phone and called my office, and I guess they'd been waiting to hear from me because they all knew I was on a plane. My mother was calling the office freaking out because she didn't know what airline I was flying. She was screaming at them to find me. I got one of my cops on the phone and said, "Cathy, what the hell is going on?' 'Well, Sarge,' she said, 'they just blew up the Twin Towers. And to make matters worse, there are four planes they still can't contact, so they're going to shoot them out of the air.'

"Finally, about 10:30 A.M., we landed in Atlanta. By then we knew that both the Towers had been hit, and somebody said that the Pentagon was hit, too. I was just happy when the damn thing landed, because I thought that four planes were going to get shot out of the air. When the plane landed, everybody clapped, and you could see tears. Nobody went hysterical, but you could see tears rolling down people's faces, and I was still sitting there in complete disbelief.

"They told us that we were allowed to use our cell phones then because we were on the tarmac, but all the cell sites were down, so I was having a hard time getting through. By then I was starting to find out who were the cops on the plane, who were the feds, who was this one, who was that one. Some people were getting through to their offices, so we were getting information on the plane, but the pilot was still not telling us anything. We sat on the tarmac for three hours.

"The guy next to me told me that his wife worked for Hertz. I said, 'Great! Call your wife and get us a car, because I've gotta get home.' We got into the terminal and he started to get in line with 400 other people and I said, 'Go to the front of the line, tell them who you are, and get us a car.' He went to the front of the line and came back 15 minutes later with no car.

"So I saw this other guy with a Marine Corps emblem tie clip. I asked, 'Where you going?' He said, 'I gotta get north.' I said, 'I gotta get north, too,' so we hooked up and hitched a ride from the Atlanta airport to the next closest town, which is in Macon County, Georgia. We got to the airport in Macon County, and it was just this little tiny runway with a bunch of twin-engine planes and that was it. The pilots were all walking around in a daze.

"There was this little 13-inch black-and-white TV sitting on a counter in the middle of the airport office, which was about as big as a hotel room. This was the first time I actually saw the Twin Towers come down, and I just started to cry. I'm a cop. I'm a New Yorker. Nothing fazes me. But this was something totally different. Not to mention the fact that I was told on the phone that half the police department was dead. I heard 300, 400, 500 cops, this one, that one, all this crazy stuff. I thought at that point that the whole upper echelon of the police department was dead, including my chief. I was completely devastated.

"Finally we were able to rent a car in Macon County, but they told us that I-95 was closed northbound. Being in such a daze, I didn't realize that my shield would have gotten me through to where I had to go, that it was gold at that point. But I wasn't thinking. So I drove almost nine hours with this Marine and some other guy back to West Palm Beach, which I didn't think was such a big deal because they told us that the airport was going to open again at 5:30 the next morning. I figured I'd be back in New York City by 9 A.M. on Wednesday.

"Well, as we all know, the airports didn't open for two days. I was glued to the TV set. I slept maybe three hours a night. I had CNN going all the time. I was calling work to find out who was alive and who was dead. I had the radio on. It was so frustrating because I felt so helpless. I wanted to be there with everybody, and I couldn't. They weren't running any trains, and at one point, I was even thinking about buying a car to drive back. Finally, on Friday, the chief called and said, 'Listen. I got you a flight. Be at the airport in two hours.' And that's what I did. One of my guys picked me up, we drove from JFK right down to Ground Zero, and I was there for three weeks.

"This was three days after it happened, and I was still walking ankle-deep in ash. I was still seeing remains. And the smoke, the smell of it, breathing it in—we were all coughing. Everyone was coughing for weeks. There was one wall that was spray-painted 'Morgue' with an arrow pointing down the block. It was surreal.

"To watch it on TV is one thing, but to actually be down there and smell it and see it and stand next to a building that's totally destroyed is unbelievable. The Twin Towers covered the area of four football fields, so you're dwarfed when you're down there. People were still walking around in disbelief. People who had been down there the entire time were still walking around, looking up, and just shaking their heads. You hear about this stuff in other parts of the world, but it's nothing you'd ever imagine here. This is New York. This is the greatest city in the world.

"They divided the downtown into sectors, and we had a sector close to Ground Zero. My chief had the financial district, so we were responsible for Wall Street on that first Monday when it reopened. We were sending cops in for crowd control, going down to the site, getting people in and out buildings, assisting business owners and store owners down at Ground Zero, securing their property, and retrieving property.

"I didn't have any days off, but on my off-duty time, I would go down there and volunteer on the bucket brigade. It was very frustrating. I was standing there, just handing bucket after bucket, and I wasn't finding anything. Ash, rubble, just crap. There were hundreds of detectives raking through every inch of this debris and rubble out at the landfill on Staten Island where they were bringing all this stuff. They were finding people's wallets, they were finding little mementos, they were finding bones. That's what they were using to identify people.

"I didn't want to leave. I didn't want time off. After about three weeks, the chief said, 'Everybody's got to take a day off. Everybody just has to go.' And I said to my people, too, 'You must take one day off a week. You can't be down here every day.'

"I'd been down at the morgue several times watching them open body bags, and the only identifiable thing I saw them take out of a body bag was a forearm. I saw somebody's forearm with a hand attached. But other than that, it just looked like a lump of tar, this big, burned—whatever it was. The medical examiners were having to stick their fingers through the remains of what once was a person, and the only way you could tell that it had been alive was that sometimes you could see intestines. They opened up the body bags and they ran their fingers through and they were pulling out chunks of glass and chunks of metal and throwing them in the garbage and then putting the body parts back in the bag. They were trying to identify the bodies for the families. There were detectives down there day after day, trying to identify these people, just so that one family could at least have a proper good-bye. It has to change your life.

"They always say that cops have the best and the worst sense of humor. We can sit and eat a slice of pizza over a DOA or the most horrific homicide scene. That's our way of dealing with it. But none of that was going on down at Ground Zero. There was no joking. There was no laughing. These guys had been down there looking at it every day. Maybe it was the first time in their lives that some of them were showing this kind of respect. The most seasoned detectives who've worked in Homicide for 20 years have told me, 'I never in my life imagined that I'd be seeing something like this.'

"There are still almost 3,000 people we haven't found. There are 3,000 people lying in a pit. There are 3,000 families, hundreds of thousands of people who knew, or loved, or lived with these people. It's the biggest grave this country's ever seen.

"The hardest challenge for me right now is to try to remain professional and not break down and cry. I'm listening to stories and listening to family members, and I'm trying to be strong for these people. I feel for them, and I'm supposed to be strong for them. The most difficult thing for me is maintaining my composure. I've seen guys in full uniform just break down and cry, and nobody cares. Nobody calls them a wimp. Nobody's making fun of them. I see very little distinction right now between men and women and feelings. This may be the first time these men have ever felt this deeply in their lives.

"I have 12 years of experience, and I've seen some pretty horrible things in my career. I just try and separate as much as I can. More often than not, the victim of a homicide has something to do with the perpetrator, so that level of detachment is understandable. But when I have that kid or that truly innocent victim of a rape or homicide, I take that home with me. That stuff never goes away, but I learn to move on. It's self-preservation.

"This is totally different. These were 3,000 innocent people whose only crime was going to work. They got up that morning, and they went to work. I mean, I've been on mass suicide scenes and homicide scenes where three or four people are killed and there's blood all over the place. I was on the scene when the city marshal was burned to death. These are horrible things. These are things that I keep with me, but I move on. It's part of my job. But not 3,000 people.

"After September 11, I didn't really go home for a month. If I did go home, I'd literally pass out for a couple of hours and then go back to work. Sometimes I slept at work because I didn't want to go home alone. It's tough. I think you've got 33,000 uniformed cops and 11,500 uniformed firefighters who are going to be screwed up for the rest of their lives.

"This is the first time I've had two days off since the incident, and I've kept myself busy. It felt good to just be home. A friend of mine came over last night and we watched a game on TV. It's hard to go to sleep at night, though. It's hard to go to sleep alone. I think about what I've seen, I think about what I've heard, and I think about the stories. It just plays over and over again in my head. I'm restless. I have pent-up energy. And as soon as I get up, I go to work. There's no time to go out and party. Nobody wants to go out and party, and I'm surprised at that. At first I thought, *Here we go. Good excuse to drink. Let's go get plotzed.* But I'm working 12, 14, or 16 hours a day, and I'm exhausted when I'm done. I'm on my feet, running around, and the stuff I'm dealing with is emotionally as well as physically draining. And then I've got to get up and go to work the next day.

"This whole incident has made me reevaluate what's important to me. I take a look at my life and ask myself, *What can I do? What can I do for these people?* I've never been so proud to be a cop. I've never had so many people come up to me, put their hand on my shoulder, shake my hand, give me a kiss or a hug, and say, 'Thank you for doing what you do.' Strangers on the street walk up to me and say, 'How you doin' officer?' and they mean it. This is not something the NYPD is used to.

"There's no place I can go right now, especially in uniform, where I'm not treated like gold. Everybody's still bending over backward for everybody else—even the CEO of Citibank. I walked up to a table and wanted a cup of coffee, and he jumped out of his chair to get it for me. Between the money and the donations and the food and the water and people bringing home-cooked meals for us and socks and shoes and hats and gloves, the unity that's come out of this is just amazing.

"I don't want people to forget how we came together as a nation and how we all stuck together through this. Whether you were black, white, green, or purple, nobody cared who you were standing next to. Even Democrats and Republicans weren't killing each other through this. We were united, and I liked seeing that. The patriotism is awesome. Seeing the flags all over the place is incredible. This country was built by immigrants, particularly in New York. There are more immigrants in New York than anywhere else in the entire country, and nobody cared who was who or what was what. Everybody was helping everybody. I think that's the most important lesson we can take from this.

"What you're doing with this book and just being able to talk to you is good. I haven't really talked about it. I get the 50 questions from everybody: 'What's going on? What's it look like? Blah, blah, blah.' I don't want to talk about it. You are the first people I've really spoken to about it, and this has been good for me.

"I have a couple close friends who are cops, and we sit down together and purposely *don't* talk about it. We look at each other and just know. There's no need to discuss it. It's there, it's done, let's do our jobs and move on. We have to go forward, forge ahead. Get back to where we were. Be even bigger and stronger than we were before. Try to do everything we can to help anybody we can.

"I had a brief moment, a very brief moment in Florida, when I thought about telling the NYPD, 'You know where my gun is, you know where my shield is; I ain't coming back.' But no. There's no way I'd walk away now. Now it's a matter of pride. It's a matter of going out there and rebuilding New York City and getting the country back to where we should be. Everybody's life changed on September 11, and we'll never be the same. But nothing is going to break our spirit."

MAJOR KALLY EASTMAN

AGE 40
DEPUTY AREA ENGINEER
U.S. ARMY CORPS OF ENGINEERS
NEW YORK DISTRICT

Major Kally Eastman has spent 20 years as an engineer for the U.S. Army Corps of Engineers. After listening to Kally talk about her engineering experience, it was no surprise to us that she was called in to Lower Manhattan within two hours of the terrorist attack to help coordinate debris-removal operations at the World Trade Center site.

As Deputy Area Engineer for the New York District of the Corps, Kally has been stationed at West Point Military Academy for two years. She had both the technical expertise and the knowledge of local resources necessary to help guide FEMA and the City of New York through unprecedented engineering challenges at Ground Zero.

"I'm a very practical and pragmatic person," she said. "My contributions may not have been as large or dramatic as others people's, but I'd like to think that I helped make the process run more smoothly."

As a troop commander and in other roles for the Corps, Kally has traveled around the world to manage heavy construction jobs and build military bases for the armed forces.

"The personal aspect of moving around so much has been very challenging in terms of establishing long-term relationships," she told us. "My average time in any one place was 28 months, so I rarely kept my address for more than 2 years. I've moved 13 times in 20 years."

Kally retired from the Army in June 2002 and plans to stay in one place for a while.

"At any given moment, I may have needed to get 15 shovels in there. I may have needed to get six experts. I may have needed to put 50 people on the ground. We had to figure out the mechanics of it, coordinate it, get it funded, and get the people to do it."

—Major Kally Eastman, November 23, 2001

"I was contacted by the U.S. Military Academy at West Point when I was in high school. It seemed interesting, so I joined the Army in 1978 and then went to the Academy for four years. When I got there as a freshman, the very first women to come into the Academy were seniors. That was the first year that women were in all four classes. I came out of West Point with a general engineering degree and was commissioned into the Army in 1983.

"There aren't a lot of women in the Army Corps of Engineers. It's a difficult branch for women, because it's still very much a he-man environment. I walk onto a project site, and the assumption is that the guy will know what he's talking about and I won't. I have 20 years' worth of experience, yet I don't get the benefit of the doubt. I understand that it also works that way for women engineers in the outside world. Women have a long history of being nurses and secretaries, but we don't have much history as engineers.

"I spent three years in Korea as Director of Engineering and Housing, which is equivalent to facility engineer or facility manager. I had a crew of 300 and oversaw base maintenance, operations, and construction. In Saudi Arabia, I was commander of Charlie Company of the 92nd Engineers Combat Heavy, a troop unit that did construction projects. I had my own heavy equipment operators, plumbers, electricians, carpenters, maintenance section, cooks, and medics—all those people in one company.

"I took a 15-month tour in Johnston Atoll, which is a little bitty island west of Hawaii. It's about the length of a runway. The military operated a plant there to deactivate chemical munitions like mustard gas. I worked on the base side, where we had to make our own water and electricity. Our food came on a barge, and produce came on a plane. I also spent 90 days in Hungary and Bosnia, where we built base camps. I was like the base camp mayor or city engineer. I managed 51 real-estate leases, negotiated requirements, and managed the work.

"After serving as Deputy District Engineer in Savannah, Georgia, and Deputy Commander in Buffalo, New York, I came back to West Point, where I'm the Deputy Area Engineer for the New York District.

"On September 11, I was at West Point, about 70 miles north of Manhattan up the Hudson River. When I heard that a plane had flown into the World Trade Center, I called the Emergency Operations Center in Manhattan and left my number. The colonel there called me back and said, 'I need you to go to the FEMA mobilization site.'

"FEMA is the Federal Emergency Management Agency. When FEMA activates in a disaster, its engineers are the Army Corps of Engineers. They bring in a specially trained cell of civilians and utilize Corps assets to help manage the emergency. The Army Corps of Engineers is a huge organization of about 35,000 people. There are about 500 or 600 active-duty military and the rest are civilians.

"I went down to the site with my cell phone and began to coordinate with headquarters to receive all these people coming in. I took a couple civilians out of the New York District, and we went down and set up in Edison, New Jersey, with the Federal Coordinating Officer, who was the presidential appointee assigned to handle the emergency. We set up in the same location, and we were their engineers on the ground.

"We started bringing in people who were specially trained in infrastructure, public assistance, buildings, utility assessment, and debris removal. I was a local person who knew what resources we had available in the New York District and in the North Atlantic Division, so I was one of the resident experts.

"We brought in the first group of guys from Louisville, Kentucky; Norfolk, Virginia; Mobile, Alabama; Vicksburg, Mississippi; and Jacksonville, Florida. The people coming in needed somebody local who could answer their questions about the area: How do the airports work? Can you get across the river on a boat? Can you get a train? Are there hotels available? Because I was a local person, I knew the answers.

"These guys made up the technical-expert group to help put missions out on the ground. They're trained to respond to natural disasters such as earthquakes, floods, and hurricanes. In those situations, you normally don't have roofs on houses, and you don't have water or utilities. You need organic debris removed, and you need equipment like wood chippers and dump trucks.

"But this was different. This was an instantaneous, human-made, mass disaster. It was a created disaster. This was not what anybody practices for. This was not what anybody prepares for. You don't plan for a 16-acre debris pile. You don't plan for these kinds of mass casualties. You don't plan for this massive amount of steel. In an earthquake, a building gets damaged, but it doesn't compress itself and pancake down like the Twin Towers did.

"These buildings performed as they were designed to perform, which was actually a benefit. If those buildings had gone in one direction or another, if they would have whiplashed or snapped, the impact area would have been huge. Who knows how many more lives would have been lost?

"We were dealing with a debris pile, but it was also a crime scene and a mass grave. In other parts of the world, people have been found alive 21 days after an earthquake, so there was a lot of hope, and it was approached very positively and very respectfully. I really believe it was handled with the proper respect.

"When you're dealing with a natural disaster such as a hurricane, it might hit 15 counties or 6 cities. All these municipalities need assistance, and no single entity is in charge. You come to New York City, and New York City's in charge. New York City has the world's fifteenth-largest army in uniform—its police department. New York City is second only to the federal government in its purchasing capability. The city had several agencies—a contracting organization, a construction organization, and a Department of Sanitation—all in place. The destruction of the World Trade Center was a bit beyond their capabilities, but they did a great job. They jumped right in and took care of things.

"We brought in debris experts, who helped do the computations for how much debris there was. They discussed with the city how to move the debris and what kind of equipment was needed. We had to figure out what would happen to this debris pile if we moved that one, or what the impact would be if we added 200 or 400 more people to the site. Those were the kinds of engineering questions we were asking. We didn't have the same kind of manageable, measurable targets that we would have had under normal circumstances.

"Four of the world's largest construction companies were onsite doing the debris clean-up. But if you went from corner to corner to corner, they were all doing it differently. Who was the most efficient, and who was the best? That's where the debris people came in and said, 'Hey, if you move this to that site, then you won't have to handle it three times.' We worked with those kinds of things.

"What we realized was that we couldn't just truck all this stuff everywhere. We decided to dredge the river, put barges in, load the debris on barges, and take it to the landfill on Staten Island. The Corps worked out the permit issues and did the dredging. We hand-walked all the permits through so that we could dredge the river to remove the debris from the site.

"Because I was local, I knew who to call if I needed an expert on dredging, permits, real estate, or anything else. I was the liaison into the District to find the expertise we needed. I went out to Governor's Island and helped do an assessment of the facilities for the National Guard. We brought in structural specialists, who helped do the mapping at Ground Zero, and I went down and looked at the pile with them to help figure out what we were going to do.

"California Urban Search and Rescue (USAR) Task Force One came with FEMA, and we put several structural specialists with them. We were the coordination cell for all the Corps of Engineers USAR resources. The city had no command sites, so the Corps of Engineers brought in mobile trailers with phone lines and set them up around the perimeter. The Corps of Engineers operated those for the fire department and search-and-rescue people.

"I was very busy. The New York District has a whole section of boats, and I was staying in New Jersey, so I'd get up at 5 in the morning and ride a boat over to Manhattan. I'd work until 7 or 8 at night and catch a boat back to New Jersey, eat, sleep, and come back. I was running 18-hour days for almost a month, and it was very tiring and stressful.

"I didn't have my hands in the pile, but they had a lot of other people doing that. Getting the right people to help seemed to be the key piece. What is it that needs to be done, and what assets and resources can we bring to the job to get it done? Here's what I've got. What do you want? It's a phone call away.

"The Army works one way, FEMA works another way, the city works a third way, and we all had to work together. Everyone was saying, 'Here's the right answer.' But everybody can't have the right answer, and everybody's answer wouldn't work. Sorting through to the proper path was challenging. How we were going to move the money, how we were going to bring in the people, where we were going to put them, and how we were going to use them—all those issues came into play.

"My biggest test was dealing with the interaction among people. There were people saying, 'This is my job, this is my mind-set, and here's what I'm going to do.' They were on a path, and it was very narrow. I was trying to get people to look at the bigger picture and not be so focused on details. The Army loves boxes and lines and definition, but this was a 'think-outside-the-box' type of event. Getting people to break their established paradigms was probably the biggest challenge. I'd say, 'You can't do things the way you've always done them. You can't use the book answer, because, in this case, the book answer doesn't work.'

"My experience in troop command really helped prepare me for this. I have had hundreds of people under my command, and I had to keep them focused on what their jobs were and what their mission was. I was responsible for performance windows, timeframes, setting reasonable expectations and goals, breaking things down into manageable portions, helping people learn to manage themselves, and teaching them how to think for themselves.

"You've got to be able to think something through and figure it out. Somebody called one day and said, 'We need three heavy-lift helicopters.' I said, 'Why do you need heavy-lift helicopters?' They said, 'Because we're going to sling-load and lift debris.' I said, 'Not a good idea. One, you've got a downdraft from the blades, which would put more dust in the air. Two, what happens if the debris is heavier than the helicopter, and you snap the helicopter down onto the pile?' So I helped filter ideas. What's the benefit to doing it; what's the drawback? We kind of war-gamed it. Field-risk analysis. The pros and cons.

"People were still very nervous. Very nervous. There was no way of knowing if this was all there was. Maybe there were 10 more attacks in the wings. So there was always the worry, *Are we next?* People who worked in other tall buildings didn't want to come back.

"I was so focused on the immediate needs that I didn't get a whole lot of the big picture. I went in and did my job. I didn't watch TV. I didn't read newspapers. I went two weeks without knowing what was really going on. I was in kind of an info-void, but I was focused on what I needed to do. *Here's the job, here's our mission, here's this disaster. We need to handle it.* That allowed me to stay focused, to be of service and of help, and to not get sidetracked into issues that were more emotional and would distract me from what I needed to be doing.

"I realized, after the fact, that I'd missed the whole event. I was part of it, but I didn't really know what had happened. I didn't know all the stuff around the outside. It was like fighting a fire. You're not worried about planning for the next one or planning for the area around it. That was not on the radar screen at the moment. I saved a bunch of magazines, and sometime I'm going to sit down and read all about it.

"One of the things that allowed me to keep a limited amount of distance was that I consciously read no lists of missing people for the first few weeks. I heard through rumors that none of my friends or anyone I knew personally was hurt either here or at the Pentagon. I went to the site, and it had a very strong impact on me. But for the most part, I could keep myself a step back.

"The only time I didn't was when I saw them recover a firefighter. I just stood there and bawled. I still do, when I think of it. It was a difficult moment. It was just as difficult to watch the people standing there as it was to watch the one being carried out. Everything stopped, and an incredible amount of respect was shown. Everyone pulled off their helmets and saluted, and I could tell it was somebody they knew.

"But I had to keep myself focused on the job in order to get it done. At any given moment, I may have needed to get 15 shovels in there. I may have needed to get six experts. I may have needed to put 50 people on the ground. I may have needed to get 10 of these or 100 of those. We had to figure out the mechanics of it, coordinate it, get it funded, and get the people to do it.

"I love being an engineer. It's a very up-tempo kind of work. I have steel-toed boots and a hard hat, and I'm absolutely fascinated by construction and engineering. I could have stayed down there all day and watched dump trucks and cranes work. I find that very interesting to watch from a scientific point of view. How do they move this piece of steel? How are they going to rig it? How are they going to cut it? I get totally absorbed watching that stuff. I was fascinated watching the scissors cut the steel to move it over to this guy, who lifted it over to this one, who loaded it on the truck. I kept thinking, *If they skip this step, they could be more efficient.* I can't always accept the way things are happening. I've got to fix it a little bit.

"When I was young, I wanted to do something that women didn't normally do. I wasn't going to be a nurse or a schoolteacher. I wasn't going down the traditional path, and my parents were very supportive. I hunted, played basketball, ran track, then joined the Army. One of the rules in our family was, if you start it, you finish it. You don't quit in the middle of anything, and you always do your best. I hated chorus, but once I joined, I couldn't quit until the end of the year.

"I wanted a tractor like my grandfather's when I was small. I still have that little metal tractor from 38 years ago. My parents were very good about letting my sisters and me do whatever we wanted to do and supporting whatever our choices were. That support is essential to getting women out of traditional roles and into careers that are more nontraditional. I have a niece who's a year-and-a-half old, and she'd rather play with trucks than dolls. That's good in my book.

"I believe that the hard sciences and engineering teach you to think, question things, and challenge things. Once you learn to think in a logical manner, the applications are endless in terms of what you can do with your life."

ROSE ARCE

AGE 37
PRODUCER
CNN
NEW YORK

As a journalist, Rose Arce is committed to bringing people the truth. On the morning of September 11, she went down to Lower Manhattan to cover the crash of a plane into the World Trade Center like any other breaking news story. It wasn't long, however, before she realized the importance of her broadcasts and how her reporting might help people who were trapped in the burning Towers.

Rose has been a journalist all of her adult life and has been recognized with numerous awards. As a newspaper reporter for New York Newsday, *she shared a Pulitzer Prize with a colleague for spot news reporting of a subway train crash at Union Square. She also won two Emmys for spot news reporting at WCBS-TV in New York.*

As a television producer for CNN, Rose generally works behind the scenes to put together stories for broadcast. The circumstances of September 11 not only forced her in front of the camera, but gave her an opportunity to cover stories about people who otherwise might not have had their stories told.

"It was a strange feeling to suddenly be in the foreground of something that everyone was watching," she said. "It made the whole event more painful and more real."

Following the terrorist attacks on New York City and Washington, D.C., Rose spent five weeks in Afghanistan as a producer and on-air correspondent for CNN. She also wrote stories about the lives of Afghan women for CNN.com.

"They were asking, 'Are people getting out from the upper floors? Is there a rescue operation?' And I'd think, *That's what we need to find out. We need to get on the air and let people know if there's an escape on the roof, because they could be listening from inside the Towers.*"

—Rose Arce, November 25, 2001

"As a television producer, I'm the person behind the scenes. I've chosen not to be the person on camera, so I'm not used to people knowing my work. I was a newspaper reporter before I went into TV, so I used to have a byline and people would talk about my stories. But since I've gone to TV, I've kind of faded into the background.

"I've made my career doing spot news, law enforcement, that kind of thing, although in recent years, I've come to do a lot more people stories—stories about politics, social issues, racism, immigration, and gay and lesbian stuff. But looking back at what I've done as a journalist, what I've been known for most is breaking news.

"I came to New York in 1986, and I can safely say that I've covered every major plane crash, train crash, disaster—every one of them, including the World Trade Center bombing in 1993. I was one of the first people on that scene, too. My job wasn't that different on September 11.

"That day I was leaving my house, which is in the West Village of Manhattan. It was Election Day, and I was assigned to cover the Democratic Primary. I walked out of my apartment thinking that I was going to vote, get coffee, and walk to work.

"I think the first plane hit either right before I walked out my door, or maybe when I got outside. Then I heard something on the radio about there being a plane crash or an explosion at the World Trade Center. I don't remember whether I heard it as I was walking out of my apartment or when I was at the deli across the street. I know that it wasn't until I was across the street that it kind of clicked in me. I called the office and said, 'I heard there's something at the World Trade Center. I'm going to head downtown.' I ran down Washington Street going south, and at that point I could see the Towers.

"There were people taking pictures and just standing there, gawking. A car came by and I knocked on the window and asked this woman, 'Could you give me a ride downtown?' I jumped in her car and we raced downtown, and I think the second plane hit around the time she was letting me out of the car.

"We were right next to an elementary school, and it was the first day of school for a lot of kids in New York City. There was a lot of noise, and it was total confusion down there. There was a little girl who was screaming to her father, screaming at the top of her lungs. 'Daddy! They're doing it on purpose!' It was such an eerie little scream.

"Everything took on this otherworldly quality then, because I realized, too, that this was on purpose. It wasn't an accident. It was an attack. Total panic had broken out, and everybody was running north as quickly as they could. I was running south.

"As this huge crowd of people was coming at me, I stopped and phoned in and ended up doing a brief phoner on CNN. They call them 'phoners' in TV, because when you don't have a camera, you put people on the phone with the anchorman and you talk back and forth. I called in and described the crowd. I remember the anchorman asking, 'Is there anybody who saw anything?' I almost laughed at him because everybody was in such a panic that I wasn't going to stop anybody to ask.

"Usually the role of producer in a breaking news story is gathering news as quickly as possible and making decisions on how to get down there and get a signal into the office through the microwave or the satellite truck. It's getting the cameraman in position, negotiating with the cops as to where you can be, holding the microphone sometimes, asking questions, and making sure the reporter's up to speed. On this day, for most of the day I had no reporter, no photographer, no sound technician, and no satellite or microwave truck. It was just me. Oddly enough, my job didn't change in the least. It was still all about gathering information and getting it back as quickly as possible.

"My goal was to get inside the World Trade Center, because that's what I'd done when there was the explosion last time. I thought I would kind of skirt the police lines and ignore the police and get inside. As I got closer to the building, I realized my cell phone wasn't working. I knew I was one of the very few reporters there yet, and they were going to need me to keep calling in.

"So I abandoned plans to get inside the building just a block or two north of the World Trade Center complex and decided I had to get in where there was a phone. I ran into this apartment building, and there was a woman there, Julie Gang, who had just picked up her kid at school. I said, 'Do you live here?' She said, 'Yes,' and I said, 'I'm from CNN. Can I go stay in your apartment so that I can make calls?'

"She said yes and told me where her apartment was. People were so unbelievably accommodating that day. I went up to the top floor and starting knocking on doors until someone opened. When this guy opened the door, I said, 'Are you Jim Huibregtse? I met your wife downstairs, and she said it was okay if I came in.'

"This apartment had enormous windows and a terrace above, and it faced the World Trade Center. The Towers were so enormous that they dwarfed everything around them and made all the other buildings look tiny. They were looming above us, and we had a clear view because we were so high. The buildings were on fire, and there was so much noise you could barely hear yourself. The noise was deafening from the sirens and people screaming as they were running.

"They had a telephone, but they'd called so many people in their panic that the battery had run out. You know how it is today—nobody has a hard-line phone anymore; they're all cordless. So I couldn't use the phone unless it was on the cradle. I had to leave it on the cradle, put it on the speakerphone, give a report, then race back to the window.

"We could clearly see people in the Tower. We could see their shadows on the other side of the glass. It was a glorious summer day. It was just absolutely beautiful outside. There was enormous sun, but there was a fire inside, so the way the light was shifting, people looked like shadows inside the building. We could see that the windows were broken, and we could see these shadows inside.

At one point, people were all crowded in the windows. Jim and Julie are both professional photographers, and Jim had taken a camera with a telephoto lens up to the terrace and was zooming in. Suddenly, I heard him scream from upstairs, 'Oh my God, people are jumping!'

"I raced over to the window, and we watched these people jumping one after the other after the other. There was a guy there who looked like he was helping push people, which led us to think that these people had decided they were going to jump. When people jumped, it seemed that they immediately realized what they'd done, because they were flailing against the air current. Their shoes were flying off.

"People would hold hands and jump together. You know, the mind plays tricks with you. Jim and I both kept saying, 'What are they jumping to?' I don't think we could completely grasp that they were jumping to their deaths. We thought, in our craziness, *Is there a net? Are they going to get on the roof?* I kept imagining that helicopters would land on the roof, but no. It went on and on and on. I don't know how many people jumped, but we saw more than a dozen at least, probably closer to two dozen.

"We had turned on the TV at some point, and I was watching it out of the corner of my eye, so I knew that the Pentagon had been hit. But I was completely and utterly focused on CNN and on doing my job. I don't think I was really processing what I was seeing, other than to occasionally look at Jim and think, *Wow, he looks pretty scared. I wonder if I look that scared?*

"I kept thinking, *How do we get a camera here?* I kept dialing and dialing, and the phones were busy all the time. How could the phones be busy at CNN? They have a zillion lines. When I did get through, there was pandemonium in the office and they sounded panicked in the control room.

"I also kept trying my cell phone over and over and over again, hoping that it was going to work so that I could leave the building and go over to the World Trade Center. I toyed with the idea, between phone calls, of leaving and going over there, and the only thing that kept me from leaving was the phone. If it had not been for the phone, there's no question that I would have been inside the building.

"I called my voicemail, and there were all these messages from people who were hearing me on TV. For the most part, they were panicked because they had relatives in the building. They were asking, 'Are people getting out from the upper floors? Is there a rescue operation?' And I'd think, *That's what we need to find out. We need to get on the air and let people know if there's an escape on the roof, because they could be listening from inside the Towers.* All those calls made me refocus on what I was doing, because I immediately realized the importance of the questions I needed to get answered.

"I was looking out the window, trying to assess the situation to see if I could answer those questions, when the first building started to buckle. It was almost like an accordion thing happened. First there was some kind of explosion around the floors where the plane had come in, and then the top started to buckle. What was awful was that people were still jumping. I was thinking that they were going to be crushed by the falling building—it wasn't occurring to me that they were dying anyway.

"As the building started to come down, it created this mushroom cloud, these sort of tumbleweeds, these big pieces of the building rolling over the tops of all the other buildings. We were maybe two blocks north of the complex, and my first reaction was that we were about to be hit. Jim decided to leave. He left his apartment, and he left me in it.

"I decided to stay. It seemed that whether I left or stayed at that point was irrelevant to my safety, and there was no good decision to make. If I stayed, I retained the ability to keep reporting, whereas if I ran into the street, there was no way I was going to be able to call the office.

"I moved as far back into the apartment as I could, because I could see this stuff coming at me. I could hear debris hitting the glass and falling on the roof. I just stood there, hoping it was going to calm down and waiting for the glass to break, which, thank God, it didn't. At that point, the phone had been charging for a while, so I decided to crawl out into the next room and call again. I was looking through the glass, and it had gone from being this glorious summer day to being totally pitch black outside. All the intense noise had stopped. There were no sirens. There was no screaming. There were no cars. It was just completely silent.

"I got through to the office and described this to the anchorman. I described how it looked like this strange version of the historic blizzard we had in New York in the early 1990s, which was the first time the city had been totally shut down for any length of time because there was so much snow on the ground. It reminded me of that. There was so much of this 'snow,' which was really all these little pieces of glass and ash and paper and concrete and people. The buildings were covered with it. It was heaped up everywhere.

"I kept saying, 'Nobody's coming from there.' There had been this sea of people running out and emergency workers running in, and now there was nothing. I looked down in the schoolyard, and there was a kid who had run out of the back of the building. I could see a teacher running through the 'snow,' grabbing this kid, and racing up the street. I saw a firefighter running north, and he was carrying the body of another firefighter.

"For the first time I was really scared, because I realized the rescue workers were running. New York firefighters don't run away from fires. I've always admired their pluck. I thought to myself, *For a New York City firefighter to run away from a fire, it must be really bad down there. I should run, too. I should get out of here.*

"I hung up the phone and gathered my stuff, and when I got outside, I was very confused about where I was. It's strange how things work. You can be clearheaded about some things, and in other ways, you just get kind of muddled. It was this bizarre sort of moonwalk thing, because I could barely breathe. The air was really thick. My eyes were stinging and burning. A lot of the powder in the air was pulverized glass, and I was immediately covered with it.

"I was almost oblivious to the second building collapsing because I was walking around in a swirl. But I could hear it, and the debris flew all over the place, and it was very hard to breathe. I met up with a cameraman, and we walked farther south and ended up on West Broadway. The World Trade Center was a complex of buildings, so you had the Twin Towers and this esplanade around it, and these other enormous buildings that in any other city would be called skyscrapers. They were completely engulfed in flames. Pieces of the Towers had fallen all over them and shaved off the front of some buildings and squashed other buildings like tin cans.

"The police and firefighters had no idea what to do. From what I've seen, rescue personnel, like journalists, aren't trained for specific situations. We're trained in generalities, because emergencies by definition are chaotic. You can't really tell somebody, 'Well, if this happens, do this.' It's more that we're trained in making judgments. This event defied all the things we'd been taught. It was so unpredictable that we had to constantly balance our own safety against the need to do our jobs.

"I kept wanting to get closer and get more video and more sound and more information, and, at the same time, I was hesitant. We'd seen some pretty horrible stuff, and we knew we were in a very dangerous situation. I pushed the cameraman to go a little farther south, and we stood there taking video of this building that was engulfed in flames. We interviewed two firefighters. One guy said that he had just come from downtown and that nobody was alive. He clearly was in a state of very deep shock. He kept saying, 'I ran up and I couldn't see anybody, and I wanted to go back because there wasn't anybody.' He was just very, very out of it.

"There was another firefighter who was wearing shorts and a T-shirt that had some burns in it, and he was clearly injured. He said that he and his ladder company had raced down from the Upper West Side, and they were trying to figure out what to do, when the building collapsed on top of them. He said he dove under the fire truck, and that the fire truck had been crushed by debris. He had managed to drag his way out on his stomach by getting rid of his clothes and his helmet and coming out in just his T-shirt and shorts into the rubble. He kept saying over and over again, 'All the guys on my truck, all the guys on my truck got crushed.' He was the only one who made it out.

"I was painfully aware that there were a lot of people dead. That was also the first sign of what was going to emerge with the firefighters in the days afterward. Here was this guy, a block and a half from this building that was totally engulfed in flames, pieces of it were falling, it was clearly about to collapse, the cops were telling people not to go in, and he kept saying, 'I gotta go back. I gotta go back. There's got to be people alive, and I'm a firefighter. I'm supposed to be in there rescuing people, and instead I'm running for my life.'

"We did a lot of shooting from different locations throughout the day. At that point, it was like those old World War II movies, where convoys of soldiers are coming into town. They were all wearing red and blue firefighter outfits, and there were truckloads of firefighters limp in the back of these trucks. It was like the infantry. Chambers Street was just a long, long convoy of emergency vehicles from places I'd never even heard of. Everybody was descending on Lower Manhattan.

"That night, you could see this huge plume of smoke. There was a roaring fire, the sky was lit up like daylight, and the city was quiet. Everywhere I walked, my feet would stick in this mud. There were places where the spotlights hit the ground, and I could see that the mud was full of blood.

"The debris was made up of the particles of everything you'd find in a typical American office. There were little tiny charred pieces of the pictures people had on their desks. Pieces of their ID tags, their Rolodexes, their faxes. There was a Smith Barney salary list, and it was all charred. There were rings and purses, and all this stuff that had blown from dozens of stories above was now confetti on the ground. Now all this confetti had mixed up with the mud and become this goop full of human remains.

"I went back to my apartment that night, and the building was completely empty. Everything in this neighborhood was empty for a whole week. The streets were closed, and I could see military vehicles racing up and down 8th Avenue, sirens on. It was like being in another country. It was so creepy and sad. Just incredibly sad.

"Outside, there was this haze on my patio, this fog, this smoke. It smelled acrid, which it did for weeks. That was really unsettling. It was silent for the first couple of days, but every time there was a noise I'd think, *Now what?* Nobody thought it was over. Plus, the panic of the authorities made it even more acute. They were all going nutso, the president was crying on TV, and suddenly there were military aircraft now, after the fact, swooping around all over the place.

"I live just uptown from the Trade Center, and every morning and evening since I moved to this neighborhood, I've taken a walk. When you hit 8th Avenue, every New Yorker turns around and looks at the World Trade Center. It's like looking at the capitol when you're in Washington, D.C., because that's the signature view of downtown. On a beautiful day like September 11, I'd see the outlines of all those buildings and these huge Towers, and I'd feel like, *Oh my God, I'm so lucky to live in New York.* Now you look downtown and there's this enormous space in the sky. The downtown area is just empty and broken and sad.

"Since then, we've covered a lot of firefighter funerals and done so many victim profiles and essays about the heroism of people. And then anthrax happened. So it's really been a constant flow of news about the tragedy. I think the thing that's most acute now is the economic depression that's come in the aftermath. The pain that people are feeling. A lot of the stories we've done recently are about people who've lost jobs, which is its own separate tragedy. Here you have folks who survived something awful and saw something awful, and now they don't have jobs. In many cases, these are people who lost very dear friends and relatives, as well as their jobs. They've got to deal with the trauma of what they've seen, as well as the trauma of not being able to support their families. They face an uncertain future.

"We've seen the city go from what was a real heyday, a real Roaring '20s kind of atmosphere over the last five years, to just—*poof.* September 11 was the true beginning of the new millennium. That's the day the century turned. I can very much see us sitting here 20 years from now telling people how it used to be before September 11, when the tech industry boomed, and the city was defined by excess and an incredible sense of wealth and safety and power and hubris.

"Bruce Nussbaum of *Business Week* magazine wrote a column on September 11 about how, before this, words like *bravery* and *heroism* were words you'd use for financial analysts and stockbrokers and people who traded big money; when the working class stiff wasn't anything much, and people were quitting very noble jobs to become web page designers; when everyone kept saying that this was going to be the 'New America.'

"In journalism, news was changing. The world was at peace. But in a moment, we're suddenly in World War II again, with that same mentality of crisis and despair. Bravery, once again, has been reassigned to the working class: the firefighter and the cop and the national guardsman and the family that sends their kid off to a foreign country to die for a cause. It's all been refocused into what to me seems an oddly old-fashioned thing. Once again, there's the fear of dissent, and there's a tremendous amount of civil control.

"I went down to Ground Zero on Thanksgiving Day, and the military was barking at the tourists not to take pictures. Being a good journalist, the first thing I thought was, *How dare you tell them they can't take a picture!* I don't like the fact that they're taking pictures either, because it's a gawkish kind of thing, but how dare the military tell them that they can't! I thought, *What gives you the right to tell people they can't take a picture in the United States of America?*

"Whatever the United States may or may not have done—and God knows we're guilty of all sorts of atrocities around the world—none of that is going to justify yet another atrocity. The tragedy of it explains people's rage. But I think it's also important to look at the situation more deeply and see that this is what war does. This is what anger and rage do, and it's not a pretty thing. People's lives have been changed indefinitely. The deaths of the 3,000 people who were killed here have affected many, many tens of thousands more whose lives will not be repaired in any amount of time because of this.

"What the families of the victims say is that they're incredibly and profoundly hurt. That's what they talk about. I don't know if they've wrapped their heads around the politics of why this happened. For them, it's as if a total stranger walked up to them in the street and smacked their grandmother in the face. One person said to me in an interview, 'What does my mother have to do with the Middle East? She's a peaceful, progressive woman who championed the rights of the poor everywhere and has certainly sympathized with the complaints that a lot of the Islamic world would have with the United States. What in God's name does she have to do with this?'

"Americans, for the most part, don't really understand what's been going on in the Middle East and the role that the United States has played. Some of that is born of the isolationism the United States created for itself through the media and the government. People don't understand what's been going on with terrorism, so I think that a lot of them were stunned by this, because they have lived lives isolated from it.

"I think most people don't give a rat's ass about what's going on outside their own little worlds. That's the one thing I've learned over and over again as a journalist. Most people don't care that much or want to know. One of the comforts of America and its prosperity is that you have the luxury of being able to isolate yourself from all the bigger stuff that's happening in the world.

"As a journalist, I've had this argument with people for years, when they would excoriate me for writing about the murder of a kid in Brooklyn. They don't want to hear about how awful it was that he got shot, and they think it's appalling that I interviewed the parents. They say, 'How could you ask people how they feel? You're exaggerating it and making it look like the city is dangerous.' I always say to them, 'The city *is* dangerous. A kid *did* die. It's an awful, terrible thing, and his parents talked to me about it because they want everyone to know it's an awful, terrible thing, and something has to be done. They're expressing their rage.'

"I've always believed that a journalist's job is to bring the truth to people—whether that truth is nice or not nice. That's what we do. We bring the awful truth to life whether people want to hear it or not.

"The only bad decision, the only way I could have screwed up covering something like the World Trade Center attack, was to have gotten myself or somebody else injured or killed. Beyond that, there's just the truth. There's just bringing the pictures and the information to people. It's what happens after that that's so much more challenging, trying to explain it and put it into context. Yeah, you're seeing all these grieving people, but what do you do about it? That's the unanswerable question.

"Every time I go to a funeral, I'm filled the same sense of rage as everyone else. I've covered tons of funerals over the course of my career. I spent a summer doing a story a day on a kid who'd been killed in New York City, because there was such a big crime spree. Planes have crashed. Firefighters have died. Cop killings, train crashes—I've been to all of them. To be an effective journalist, I've learned to keep a huge distance from all of that.

"But this is different. This is very, very different. I feel much closer to the people who died, even though I didn't really know any of them. In essence, New York is a city of being friends with lots of strangers. This happened in my own backyard.

"My life has always been predominantly about being a journalist. Most of my close friends are journalists, too. I love journalism. I really consider it a calling. I've had a passion for it since I was a kid, and it's completely taken over my life. It's also made my life a lot sadder. There's something about this story that is a lot more shocking and a lot sadder. It doesn't get any worse than this.

"In many ways, I'm much more focused as a journalist than I've ever been, which is a good thing. I'm more committed to being a journalist, because I've had a very painful and dramatic experience that demonstrates the importance of what we do. For most of my career, I've fought against the bad things that journalism can be and that journalism does. I choose to work in the mainstream media because I want to make it better.

"CNN has done many stories about the people who've died. I know that if it weren't for me and other people who share my values, many of these people's stories wouldn't have been told because there isn't a lot of importance placed on them. We've done stories about undocumented workers, people of color, women, and gay people. We did a story about Mark Bingham, a gay man who was one of the heroes on the plane that crashed in Pennsylvania, which to do on CNN is very significant.

"So in that sense, it's given me a renewed commitment. But it's also left me with lingering fears and nightmares. This is definitely the most affected I've ever been by a story. Every time I hear the rumble of a plane, I look up and measure its distance from the ground. Every time I hear a siren, I think, *Here we go again.* It's no different than what any journalist who's been in any war zone is subjected to.

"I'm much more on edge. I plan in my head, and I'm ready for anything. If 10 minutes from now a plane flew into the Empire State Building or a bomb blew up Grand Central Station, I would be 100 percent ready to go. I really feel that it could happen at any moment, because it did. The unthinkable did happen, and now it's not so unthinkable anymore."

POLICE
DEPARTMENT
CITY OF NEW YORK

DETECTIVE JENNIFER ABRAMOWITZ

AGE 37
DETECTIVE
NEW YORK CITY POLICE DEPARTMENT
CHILD ABUSE SQUAD
BROOKLYN

Detective Jennifer Abramowitz arrived for her interview in the classic detective's trench coat, its tails flapping in the wind of a blustery New York morning. She was all business, it would seem, until she took off her coat and began describing the nightmarish details of her assignment at the Fresh Kills Landfill on Staten Island. It wasn't long before her emotions broke through her professional demeanor.

A cop for more than a decade and a half, there isn't much that Jennifer hadn't seen before September 11. Now, after raking through what's left of the World Trade Center looking for body parts and personal belongings for days on end, she says she's seen it all.

Jennifer's regular detail is with the Child Abuse Squad in Brooklyn, where children of sex crimes and physical abuse often don't even realize that she's a police officer.

"Sometimes I don't even remember I'm the police myself," she said, "because really, it's like social work. I'm dealing with a lot of very little kids. The other day I had a three-year-old. Try to get a statement out of a three-year-old. He's so busy running all over the place, and all he wants to do is throw pillows at me. How do you sit him down?"

During our conversation with Jennifer, we all voiced concerns about what long-term effects the detail at Fresh Kills might have on her psyche. Regardless of how difficult it can be to deal with sex crimes against small children, she told us she's anxious to get back to her regular job and put the horrors of Fresh Kills behind her.

"I walk through these piles, through all this gray dirt, and it's almost like I'm walking through *them*. The people. It's all that's left of them."

—Detective Jennifer Abramowitz, October 31, 2001

"I started working as a cop 15 years ago, the only female on midnight patrol in East Brooklyn. It was a really bad place. That was when the crack epidemic first started, and our precinct was the highest in crime and the highest in homicides. Shots were fired every night.

"I was taking guns away from people, and because of the number of gun arrests I made, I was sent to the Street Crime Unit, which is a citywide plain-clothes unit. Again, I was the only female in that unit. I made a really nice bribery arrest of a guy who had a lot of drugs and a gun on him, and after that, they put me in an Investigative Unit. That's where I made detective.

"I've been a detective almost six years. They sent me to the Special Victims Unit, and now I'm in the Child Abuse Squad, working sex crimes and physical abuse against children under 10. The workload is tremendous.

"Our office is right under the Brooklyn Bridge, across the water from Lower Manhattan, and we had the most beautiful view of that skyline. The Towers were right there. On the morning of September 11, we went to work and turned on the TV and saw that one of the Towers was on fire, so we ran outside to look.

"That's when I saw the second plane hit. It made a sound like I've never heard before in my life. Not only did I hear it, I felt it. It was almost like I could see the shock waves coming. We all jumped, the buildings shook, and we saw the windows kind of buckle everywhere. Then I saw the flames blowing down as the jet fuel came pouring out.

"On the street, I've been shot at so many times that it became like a game for me. It was like a high. Someone would shoot a gun, and I'd run right at him. It didn't even bother me. But I'll tell you, when that plane hit, there was nothing so scary in my life. My heart was pounding—pounding out of my chest. For at least an hour I couldn't get my heart rate back down. What was coming next? What else was in store? Now I was waiting for them to start dropping bombs.

"Of course, we got mobilized right away, which means they called us all up and said, 'Put your uniforms on, get ready, let's go.' Everyone was called into duty.

"I'm supposed to have my uniform in my locker, and it probably saved my life that I didn't. I'm the type of person who would run right into that burning building. I've always done it. I used to get yelled at by the firefighters all the time. 'What are you doing running in there? What are you, stupid? You'll get yourself killed!' I used to be like, 'You firefighters. What do you know?'

"We were right there at the foot of the bridge, and it would have taken me five minutes to run across and get myself killed. But my uniform was at home. I was thinking, *I'm in big trouble now. The world is coming to an end, and I don't have my uniform.* So I jumped in my car and flew home. It took me about 15 minutes to get home and another 20 minutes to come back, and no sooner did I get back than the buildings crumbled.

"They sent us to an area where all the 911 operators are located, and hundreds of cops met there to receive assignments. I remember the first time I heard the fighter jets coming. By then we knew that they had shut all the air traffic down, and we could hear these fighters jets coming in, and it's a very distinctive sound. We were all looking at each other and going, 'Hope this is ours, hope this is ours.' Here we were, expecting another attack, for God's sake. There aren't even any words to describe that feeling. Then when we saw that the planes were ours, I felt so much relief.

"The detectives were sent out to the hospitals to try to get people's names, to try and identify the injured. I ended up going over to Lutheran Hospital, and they were pulling victims off the ferry. I think these were the first injured people they were taking out of Battery Park. It was just a quick interview, like 'What's your name, do you know who you are?' Some of these people were very altered. I was trying to find out who they were, who I could call for them, what building they'd been in, stuff like that. They were in shock, real shock. They couldn't even talk. They didn't know what to say, didn't know what to do, didn't even understand what had happened.

"They were bringing in a few firefighters and EMS workers, too, and these people were great. They were barely walking out of the rubble and they were insisting on going back. 'Just send me back,' they'd say. 'Fix whatever it is quick, 'cause I've got to go back. Give me a drink of water, give me some oxygen, whatever it takes, because I've got to go back, got to go back.' They wouldn't listen to anything else.

"The hospitals were all set up. Everybody came down. Even if they were off, people knew there was a job to be done, or were hoping there was. They had all kinds of triage areas set up, beds set up, wheelchairs lined up on the sidewalk, big buckets filled with bottles of ice water. Everyone was waiting. They had everything ready to go. They had a whole triage system in play, but the patients never came. The reality of it was, we finally realized, that if the patients weren't coming, it was because there was no one left to come. People got sent home eventually, one by one, and they started leaving, cleaning up the sidewalk, taking away all the equipment.

"There was great fear of retaliation starting in the streets, especially in Brooklyn, where there are a lot of mosques, and there was concern that things would get out of hand. They sent all the patrol officers out, and down at the Towers it was mostly fire personnel.

"They eventually sent the detectives to the Fresh Kills Landfill on Staten Island, where they were taking the material recovered from the World Trade Center. We were all detectives up there, the NYPD, the FBI, and FEMA, all looking for evidence. At first, they wanted us to find those black boxes from the planes, but that never happened. Now it's about recovery. I'm still going out there two, three times a week. That's 12 hours of manual labor. It's hard work. It's hard physically, and it's hard mentally. We're looking for anything, any part of a person, to give the families some kind of closure.

"There was a lot of chaos out there in the beginning. I put all this protective stuff on and I looked like a space monster. In the beginning, dump trucks would come right from Ground Zero—huge trucks—and they'd just dump stuff out. Then maybe a payloader would come by and try to spread it out. There were huge pieces of twisted metal everywhere. We'd take a rake or a shovel and move through the pieces and try to separate them, looking for whatever we could find. Then they would bring in the dogs, and the dogs would go through to make sure we hadn't missed any body parts. Those poor dogs. They were so frustrated because they wanted to find someone alive and all they were finding were pieces.

"We found part of a police officer's shield one day. We found a couple of guns. One of them was actually in a hand. We found a lot of pieces of uniforms—firefighters' uniforms, police uniforms, firefighters' masks. I found people's skin, and it looked like a piece of rubber. I picked it up, I looked at it, and I thought, *What's this?* I've found lots of pieces, lots of parts, like a shoulder, a hip, fingers, and scalps. There were a lot of explosions through the elevator shafts, and people were blown right out of their clothes. I found their clothes, and they were just shredded. I found their sneakers. There were shoes everywhere.

"I walk through these piles, through all this gray dirt, and it's almost like I'm walking through *them.* The people. I know it's all that's left of them.

"It's sad. I'm picking up their stuff, and I know they're here. I can only hope it happened so fast that they had no knowledge of it, that it was quick, that one minute they're running, and the next minute, poof, they're gone, and there's nothing left but smoke. Some of what the dump trucks pick up is cement and drywall and parts of the building, but it's people, too. They're just part of all that now, this gray dust everywhere. Anyone who was up high, who was in the crush of the collapsing buildings, there's just absolutely nothing left of them.

"I look at all this stuff that's completely charred and burned up and gone, and then I find something that's totally intact. And I wonder, *How did this make it with just maybe one little black mark on it?* I found two Port Authority unit citations. They were in little glass cubes, and in the center there was a floating coin. I guess it was some kind of award they won that they kept in their offices. I take that stuff and put it in a bucket. There's a bucket for personal effects and a bucket for body parts. That's all that's left.

"One day they dumped a load of fish. I don't know where they got these fish, maybe from a restaurant that was up there, but a whole truckload of fish. Then we had to go through the fish to try and find people. And the smell. As if the smell up there weren't bad enough, now there's this whole pile of rotting, disgusting fish. Another woman and I were the last ones standing in this fish pile trying to do it. We didn't like the smell any more than anyone else, but all the guys had left. I looked at her and said, 'Are we stupid? Why are we the only ones in this big old fish pile?' There was obviously nothing else there but fish—big fish!—swordfish and these big red round fish.

"It was really hard up there in the beginning because there was no organization and stuff was just thrown. I knew there was asbestos. The smell up there was—ugh—it was disgusting. Mostly it was the methane gas from the decaying garbage. The ground bubbled, and they kept spraying it with water to keep the dust down. I was looking down, and the ground was bubbling around my feet. I was thinking, *This can't be good.*

"The respirator I was wearing was given to me in the beginning, and they said that after a week the filters should be changed. So I asked the guy for some new filters, and when he gave them to me, I looked at them and said, 'Wow, these are much bigger than the other ones. Are they going to fit?' And he said, 'I guess no one told you that you've had the wrong filters on all this time.' So for over a week I was breathing in asbestos and all this other stuff, because obviously the respirator I had wasn't stopping anything.

"One night I was so exhausted. It was about 1:30 in the morning, and I was freezing. I was sitting there on the wheel well of a portable generator that was running the lights, I was wearing my whole get-up, and I had my rake in my hand. I was leaning on it with head down, my respirator and my helmet on, and I was falling asleep. It was pouring rain, and here I was sleeping on my pole, with the monsoons coming around me. The energy was completely drained out of me.

"The suits are disposable, and we get new suits all the time. I have to take it off every time I go in to eat, because I can't go into the eating area with it on. One day it was pouring rain, and I had to get this thing off. I was waiting in this tremendous line to get food, it was pouring rain, I was getting soaking wet, and I was miserable. Here came the priest from the Army out there saying prayers, and because I was starving, I was like, 'Listen, pal, if you don't have some food for me and some dry clothes, I don't care about God right now.' I get so famished after a while because I'm working so hard. I just want to eat and it's raining. It seemed like it was raining every time I went out there.

"I guess we all had trouble sleeping. I kept having nightmares, thinking, *What's next?* I was always thinking of chemical warfare and wondering what they were going to drop on us. It was like being on the edge of my seat, waiting for the next thing to happen. A lot of people were drinking to go to sleep. I'd hear people in my office who don't normally drink say, 'I had to go home and have a couple glasses of this, or a couple shots of that, just go to sleep.' I think that was pretty widespread. It's hard to go home and go to sleep after a 12-hour day, which is really like 14 or 15 hours once you get done. Then you have to get right back up again. I couldn't even keep my dog at home. I had to ask a friend of mine to keep her for three weeks.

"They say it's going to take at least a year to go through everything at the dump, but they've gotten very organized now. They've got tents set up, they've got heated locations, and now they wash you down, they wash your car down, and they're taking all these safety and health precautions. They've got these big conveyor belts, and the stuff is put on those to go through sifters. It goes through four or five different places, it's looked at and relooked at and relooked at. Then they spray-paint each pile to show that it's been searched.

"I think things are starting to get into more of a routine. We're working normal hours again, and I'm back doing some of my regular cases. There are more people rotating shifts at the dump, and they're breaking it down and saying, 'No, I don't want you going this many times. Someone else will go.' They're spreading the duty around, and they're making those who really didn't want to go out there take their turns. Now I'm only going out there a few times a week.

"This is still very hard on me, because it's not just going to a car accident and seeing somebody hurt. This is such a tragedy, knowing that there are thousands of people we're never going to find. It's especially hard when I pick up personal things that have survived, like a family picture that was on somebody's desk. I found a few women's pocketbooks—what's left of them—some with their IDs or their kids' pictures. I pick these things up, and I know that these lives have been shattered and torn apart. I try not to think about it, but it's hard not to.

"It got to the point where I was thinking about it too much. I didn't even want to do any of the usual things I enjoy. Here I was, training for my black belt test, and I didn't want to go. I was glued to the TV set when I was home, and my mind was so preoccupied. The stress even screwed up my menstrual cycle.

"Finally I said to myself, *This is ridiculous. What am I doing?* And that was it. I haven't watched TV since, I haven't read the newspaper, I just *whoosh,* go right to the crossword puzzle. I didn't want to know any more, I didn't want to see the people's faces, I didn't want to read their stories. Every day I was picking up the pieces of their lives, and I didn't want to put this piece of skin with that story in the paper. I needed to just do my job, and that was it.

"I've put myself back in training for my black belt, and when I'm there, I'm very focused. It's a great way to take my mind off everything and go into another kind of zone. I'm trying to move past it and put my life back together as normal as it possibly can be, considering there's something new going on every day and I wonder what I'm going to have to do next.

"Last week, I was detailed to 39th Street and Broadway, which is a few blocks up from Ground Zero, standing on a corner for eight hours waiting for an anti-police rally to come past me. It was amazing how many people came up to me and started conversations. Everybody who walked by was like, 'Hello, hi, how are you? You okay? Can I get you anything?' People at the bank said, 'If you want anything, just come inside. Would you like me to go get you a cup a coffee?' Never, never in my 15 years as a cop have I ever experienced this. People used to walk by me and sneer.

"People asked me what I was doing standing there, and I said, 'There's an anti-police rally coming,' and they were like, 'No! You're kidding!' They thought it was a joke. They just couldn't understand how in poor taste that was. They'd say, 'How could anyone even want to complain about the police? You've done such a wonderful job.'

"The rally did eventually materialize. They were given a permit from the mayor's office for 300 to 500 people, and only about 50 showed up. But it was worth going out there, because I got to stand in the middle of Manhattan and see a side of the city I've never seen before. It was a wonderful feeling.

"It's such a shame that something like this had to happen before people realize that they should try to live each day. Why complain about everything? Why walk through the streets miserable, hating the world, hating everybody else? I think I've learned to appreciate life more. I see this sunrise, I appreciate this day for what it is, and I'm glad I'm here. We should feel lucky we're alive, because we don't know when this life will be taken away from us.

"There are 3,000 people who didn't know that September 11 was going to be their last day. I thought it was going to be mine, too. Even when I went back home to get my uniform that morning, I called my parents' house and left a message telling them how much I love them. I said good-bye to my dog. I hugged her and said, 'I love you,' because really, I didn't think I was coming back."

MOTOROLA

CAPTAIN ROCHELLE "ROCKY" JONES

AGE 43
CAPTAIN
FIRE DEPARTMENT OF NEW YORK
ENGINE 4
LOWER MANHATTAN

It was difficult to keep up with Captain Rocky Jones during our interview, because she talked so fast and moved so quickly from one idea to the next. It was obvious to us as we listened to her thought process that she is skilled at assessing situations and making decisions at a very rapid pace.

Rocky is the highest-ranking woman in the Fire Department of New York. She was born and raised in Brooklyn, where she spent much of her career as one of the first female firefighters in New York City. She made lieutenant in 1994, and in 1999 she became the first woman in the FDNY to be promoted to captain.

Since the collapse of the World Trade Center on September 11, Rocky's responsibilities as an officer have increased markedly with the loss of so many members of her house. During our interview, she admitted that the added roles of counselor, eulogist, and family liaison have taken a physical and emotional toll on her.

"I'm very numb," she said as our interview came to a close. "Today is probably the most crying I've done since this happened. I don't cry with the guys in the firehouse, and I don't cry with the other women firefighters. But Lois Mungay warned me that I would cry here today. She said that she had cried, and she's such a tough little piece of work, you wouldn't expect it of her. I guess it's easier to cry with total strangers than to cry with people you know."

"Every time a stranger pulls up in front of their house, they think that this is the word. Every time they read in the newspaper that a firefighter's been found, they call the firehouse and ask, 'Is he from our firehouse? Is he my loved one?'"

—Captain Rocky Jones, November 1, 2001

"In 1977, they allowed women to apply to the Fire Department of New York for the first time. Before that, if you checked 'female' on the application, it just got thrown out. This was the 1970s, and we were all young and up to the challenge. My father was a firefighter, so I knew something about it, and all the guys I hung out with in middle-class Brooklyn were paying somebody to help them prepare to take the written test. I thought that was the biggest waste of money I'd ever heard of in my life.

"I wound up having a bet with these guys that I could take the written test and score within five points of their marks. I wrote as high as they did, but the physical was much more difficult. We had to do a one-arm bent-elbow hang, do a standing broad jump, and climb an eight-foot wall. I mean, the fire department has ladders. Why would you climb an eight-foot wall when you could just toss a ladder up there? It seemed that they went out of their way to find things that women were not good at.

"None of the women passed the official physical that year, and a class-action lawsuit was filed to get a nondiscriminatory physical for the fire department. After five years we won the lawsuit, and they implemented tests that actually related to firefighting. Then it became dragging hose, raising ladders, and forcing doors. Almost 50 women passed.

"The majority of the women who are on the job with the FDNY today came on together in September 1982. After the academy, we were assigned to houses, and my first assignment was two years in hell. It was ugly. I was so miserable; I thought I would never stay my full 20 years. Things got better when I transferred to Engine 159 in Staten Island, and from there I got promoted to lieutenant, went back to Brooklyn, and wound up being assigned to Ladder 159. I was promoted to captain in June 1999, and I took Engine 4 a year ago.

"It's been a good experience at Engine 4. As captain, I'm the on-duty officer, so I'm in charge of the crew. I make all the decisions about how we're going to handle medical calls, fires, and other emergencies. The engine captain in the house is also responsible for the building, so I'm like the landlord on top of everything else. We have 23 people in our company.

"I've never had problems with the guys as far as being an officer goes. When I walked into Engine 4, no one seemed to care or even notice that I was a woman. It never seemed to arise as an issue. From the time I was promoted to lieutenant, I've never had problems with any of the members who were working with me. I've never had disrespect from the guys simply because I'm a woman.

"On September 11, I was in New Mexico. My husband and I had built a house, and he'd been there since July doing the finishing work. I went out there the last week of August for a month's vacation. That morning, the phone rang at 7 A.M. I could hear my husband on the phone, and I could tell from his voice that something was very wrong. I came down the stairs and he was going to the TV. He said, 'A plane just hit the World Trade Center.'

"So we turned on the TV and sat down and watched the events unfold like millions of other Americans. At that point, just the North Tower was on fire. My husband's retired from the FDNY, and he said, 'I can't believe you're missing the fire of your career. This is the kind of fire people talk about for the rest of their lives.'

"Then we saw the second plane hit the South Tower. The entire plane entered the building, and now both Towers were burning. We were glued to the TV. We knew that the staircases and elevators would be destroyed, and we were discussing how there was really no way out from those upper floors. We heard that people were jumping.

"When the South Tower went down, I knew that my people from Engine 4 and the crew of Ladder 15, which is also housed in our quarters, were in the North Tower. We're on first assignment for anything that happens at the World Trade Center, because our firehouse is only 13 blocks away. The two companies that face the Trade Center, Engine 10 and Ladder 10, are literally across the street. I knew that they would have been first due, and we would have probably come in third.

"When the South Tower came down, I turned to my husband and said, 'There are dozens and dozens of firefighters in the street who are dead right now.' From the angle of the picture on TV, I just thought the top of it had weakened and toppled. We couldn't see that the whole Tower had actually gone down. I thought immediately that all those guys in the street, all those chauffeurs standing with the rigs, every company that was coming in, and the people who were starting to enter those buildings were all dead. And then, when the second Tower went down, I turned to my husband and said, 'Everyone I work with is dead. Everyone who went there is now dead.'

"I started to cry. Tears were rolling down my face the entire day. After about four hours, my husband said, 'Let's get out of here. You've got to get away from the TV, or you're going to drive yourself insane.' So we went into town to have breakfast, and everyone in the place was talking about it. Tears were running down my face.

"I had to decide what I was going to do. My husband was going to take me to El Paso, Texas, to get on a plane, but all the airports were closed. He was adamantly against my driving home that day, which was probably a good idea because I wasn't in any emotional condition to drive back to New York. I was calling, calling, calling, but I couldn't get through on the telephones. About 4 P.M. New Mexico time, which was 6 P.M. in New York, I finally got through to the firehouse and spoke to one of the firefighters. I asked, 'How do we stand? Who'd we lose?' He goes, 'Well, I'll tell you who I've seen alive.' And he gave me three names.

"I got off the phone and said to my husband, 'It looks like out of 11 people working, we're missing 8.' Later that night I called the firehouse again, and I asked the firefighter who answered, 'Do we have a list of names yet?' He said, 'Yeah,' and he started reading off the names. When he got to eight, he kept going. I said, 'Wait a minute. We only lost eight, right?' And he said, 'No, Cap. We lost 14.'

"There were 11 people assigned to work that day, and we ended up losing 14. It was a change of tour, around a quarter to nine in the morning, and guys who'd gotten relieved and were going home stayed on. We had two probies in their last week of a 14-week rotation in our firehouse. One was coming on duty and one was going off, and the one going off said, 'I'm not missing my first fire,' and he jumped on the rig.

"Artie Barry was on vacation and loved the Southwest. He was supposed to be on a motorcycle rally in southern Colorado and was going to go from there to come visit us in New Mexico. Two days before he was to leave, he told me he wasn't coming because his motorcycle wasn't running right and he didn't want to make such a long trip on the bike. Artie and Eric Olsen were best friends. Eric had been relieved from Ladder 15 that morning, and Artie came down to meet him to do some work on his motorcycle. They both wound up going down there, and both of them were killed.

"So we lost 14 from our house out of about 45 people—6 from Engine 4 and 8 from Ladder 15. Now I knew their names. I knew that the captain who was filling in for me while I was on vacation was missing—that would have been my tour. And the lieutenant from Ladder 15, Joe Leavey, who was my best friend at the firehouse, was missing. I kept thinking about his wife and kids.

"I couldn't sleep at all that night, and the next morning my husband asked me not to leave. He's been in rescues his whole career. He's been to plane crashes and building collapses and has pulled out more dead firefighters than you could count. He said, 'Don't go back, because you're going to see things you'll never want to remember.' But we both knew I had to go back.

"I made it to New York City in 43 hours straight. I slept for about an hour and a half in a rest area at some point, and I called the firehouse pretty regularly from the road. Driving through Texas, New Mexico, and Oklahoma, there's nothing around for miles. I was on barren roadways. But once I got into Missouri and started to hit the more populated areas, every single overpass had a homemade sign on it saying, 'God Bless America,' or 'We Love New York,' or a flag draped over the railing.

"Every place I stopped for food, there was a jar on the counter where the high school was collecting money for the victims. All the road construction signs in Pennsylvania were flashing, 'Merge Left. God Bless America. Merge Left. God Bless America.' Then it dawned on me that this wasn't just a fire department thing. This wasn't just a New York thing. Everyone cared about this.

"I got to New Jersey at 5 in the morning, and I could see the site from there. I could see that the Towers were gone, which was totally surreal, because those Towers had been visible from all over. It almost looked like the sun was rising over Lower Manhattan, because all the smoke and dust were lit up with work lights.

"When I got to the firehouse and saw everybody, there was a whole bunch of hugging, a whole bunch of 'Really glad to see you,' going on, and everyone really meant it. There was some crying. A brand-new probie was at the firehouse, and I was thinking, *This poor probie. This is your first experience in a firehouse? Coming to a house that's lost 14 people?* I couldn't imagine what was going through his head.

"One of the guys, when he hugged me, just broke down crying. My husband had warned me, and I knew it anyway, that when I got to the firehouse, I was going to be mother, counselor, wife, girlfriend, and boss. I was going to be everything for quite some time. I was doing really well until that guy started crying on me, and then I got weepy. I was trying to hold it together, trying to be the strong one.

"Another guy said, 'Cap, it is so good to see you. I want to thank you for coming back.' As if I wasn't going to come back. I don't know what he was thinking. I said, 'Well, I had to come back.' About an hour later he came up to me and said, 'Cap, I'm really glad you came back. Somehow it's like Mom is here, and things are going to get better.' I said, 'Hey. Don't put that on me. That's a really big hat, and I can't make this better.'

"Both of our rigs had been destroyed, so other companies were relocated to our firehouse to take our runs. A Brooklyn engine company came, and I thought, *This is horrible.* We had all these strangers in the firehouse, on top of everything else. We had people there who couldn't possibly imagine what we were going through. They wanted to do the right thing, but there was nothing they could do to make us feel better. They basically just stayed out of our way. How uncomfortable was that for them?

"The truck that got relocated to us was my old company, which was really sweet for me. They were strangers to everybody else, but they weren't strangers to me. As they got off the rig, they came over to me right away and said, 'How you doin'?' and I just cried like a baby. I can be strong for other people, but when they're all looking at me like I'm the victim, it's really hard. So I cried like a baby on every single one of their shoulders—or more like the middle of their chests, because they were all giants working that day. It was nice to have them there.

"Every 24 hours they relocated different companies to our firehouse, which was another nightmare because we had people staying with us from all these different places. We had a guy from California and two guys from Chicago with cadaver-sniffing dogs, and there were people all over. We have a basketball court upstairs, and that turned into an American Red Cross flophouse with cots set up for people to sleep. It was just bizarre. Everything was crazy.

"Because we didn't have rigs, we weren't being officially assigned to go down to the site. A van had been loaned to us so that we could transport our people back and forth, and we were just going over on our own. One crew would go over for half the shift, then they'd come back and the other crew would go over.

"That night, we were going to try to dig where the North Tower had been, where we thought the stairwell would be, because that's where we knew our guys had been working. It was all dust and steel I-beams. I was thinking, *There are 3,000 people in here, and I can't even find a desk. I can't even find a chair. There's nothing recognizable.* I did see a piece of drapery, and that was the only thing in seven hours that I could identify. There were 3,000 bodies in there, and nothing to be seen. Just nothing. Still, we were thinking that we could dig and find something. We all still had hope in our hearts that we were going to find somebody alive.

“We were trying to find a place to dig that we thought would help. I found a battalion chief and said, ‘You’ve got to let us do something. We’ve got to feel useful.’ He told us to stand and watch as these cranes picked up bucket loads of material and dropped it into a dumpster on a truck, to see if we saw any human remains. That was our job. After about an hour I said, ‘You know what? This isn’t good for us. This is not healthy, and we’re not going to do this anymore.’ And we went back to quarters.

“Some families were coming in right away. One of our first families to come in was Joe Leavey’s family. His wife, Carole, called and said, ‘Would it be okay if we come in?’ Her 16-year-old son, Brian, wanted to be someplace where his dad had last been. They’d started out at the first house where Joe worked as a probie, then they went to a house where he’d transferred as a firefighter, and then they came down to us. That was the first family I really spoke to. That was on the Sunday after it happened.

“We all felt so much better after seeing them. His daughter, Caitlin, who’s 11, was still giggling, she was still upbeat, and she was still excited to be in the firehouse. Not that she didn’t understand that this was serious, but she was an 11-year-old girl. She wanted to see what we’d done to the basketball court because she’d heard that guys were sleeping up there. She was just a curious child.

“It was good to see them. It made us feel better, because we realized that this was something we didn’t have to fear. I think everyone was afraid to have families come up to the firehouse until then. We’re still dealing a lot with the families, which we don’t mind doing. Anything they want is right, as far as I’m concerned. I’ve been helping people plan memorial services; I’ve been a grief counselor. Some of them are taking it very hard.

“Since this happened, everything has just been a blur. A real blur. We lost 14, and now we’re missing 13. We found some of Joe Leavey. We’re kind of hoping that we’re going to find the rest of Joe.

“The memorials are upsetting and sad, and it feels like a funeral, but there’s no casket, there’s no burial, and when we leave that day, that family goes home and nothing has changed for them. They’re still waiting for their phone to ring to say that we found him. Every time a stranger pulls up in a car in front of their house, they think that this is the word. Every time they read in the newspaper that a firefighter’s been found, they call the firehouse and ask, ‘Is he from our firehouse? Is he my loved one?’

“So the memorial service does not relieve the family of any burden at all. We’re at the point now that when we find a body, it’s really good news. How sick is that? That's all we're looking forward to now. We're glad we can give Carole some of Joe. It's totally sick. But it's good for the family, because they're not waiting for that phone call anymore.

"I don't think there's any such thing as 'closure.' I hear that word 10,000 times a day. It's not, 'Okay, we've buried him; now we have closure.' It's not. That's not happening with these families. I don't know if it ever will happen in any real sense of the word. But Carole will have a coffin. Carole will have something to bury. And she won't be waiting for those phone calls anymore. She's the luckiest one of those 14 families right now because she has something to bury.

"We found Joe, and we're hoping we're going to find some more of our guys. That's going to be tough. Now I'm afraid to ask the details about who they found, how they found them, or where they found them. When they found Joe, we thought, *Okay, tomorrow we'll get another phone call saying they found the rest of the guys.* Well, that was last Saturday, and today is Thursday, so that didn't happen. Now what? I don't know if it's worse to get it over with all at once, or to drag it out. I think this is worse. Some of the families have already had memorial services, and they'll have to go through it all over again with a funeral.

"I had to do my first eulogy for our youngest member, Tommy Schoales. That brought a whole new level of responsibility. It was really hard, because I only knew Tommy for nine months, and he was a quiet kid who did all the right things. There was no way I could come up with a funny story about how Tommy did this embarrassing thing, because Tommy never did anything embarrassing. He was just a really good kid.

"So I talked about how lucky his family and friends were that they had all the years they had with Tommy—growing up with Tommy, seeing Tommy as part of their wedding parties, Tommy being godfather to their children. I told them, 'Thank you for sharing this wonderful kid. Thank you for sharing Tommy with Engine 4, for sharing Tommy with me, and for letting him be part of our lives.'

"Every family is different. There are 14 different ways of dealing with this that I've been witness to. This two-day cruise came through, and I had to find out within 24 hours if the families wanted to go. I'd say, 'I know it's a bad time to call you, but there's this cruise, and I have to let you know it's here if you want it.' Everyone reacts differently. One person's telling me, 'Well, I need six people to go.' The next person's telling me, 'How can anyone go? How can they leave without having their loved one back?'

"One guy is waiting three years for his son. He doesn't care how long it takes; he's not doing a memorial service. So every time I get on the phone I have to think, *Okay, who am I talking to? What's this family's take on this before I say anything that's going to upset them?*

"We have three great guys in the firehouse who are working with the families. They're working seven days a week, I don't know how many hours a day, and they're certainly not getting paid for all the time they're putting in. They do everything. They take the families places they need to go, help them get emergency money from the Red Cross, and file the death certificates. Now they're talking about financial planning and doing things like fixing credit cards that weren't paid because people just weren't thinking straight. The families have come to trust us so much.

"I'm afraid for these guys, afraid that they're going to get burned out, that things will become too much for them to bear. They want to be back at the firehouse, but they don't want to give up their attachment to the families. One family had a bucket in their son's room collecting rainwater because there was a leak in the roof. Unless you go into the house and you're actually there, you're not going to find that out by accident. They saw it, and they got together a crew and went out and fixed it.

"After the Oklahoma City bombing, there were something like 20 rescuers who committed suicide. I'm very concerned about that. I think this whole thing is dire for us, and I fear for the future as far as the toll it's going to take on individuals. No one has any clue what the long-term effects are going to be on people. No one's ever lost as many firefighters. We lost 343 in one day.

"There are firefighters coming in from all over the place just to go to our services, to help fill in, because our services only average about 200 people now, when we used to have 8,000 or 9,000 people show up. There are 10, sometimes 15—I think the most we had were 23 services in one day. So you can't possibly go to them all. Besides those for my own firefighters, I'm only going to services for really close friends. I would have gone to anyone's before this. If someone died in the line of duty, I just went. Now I've got to pick and choose what I can do, what I have time for, what I'm emotionally and physically able to handle. It's very difficult.

"When it's friends I've known for years, I want to go and I want to be there, but they're also the harder ones to go to, besides for my own people. I'm still dealing with all the stuff I did before—all the administrative work, all the fire duty, all the decisions—plus dealing with families, counseling the guys, watching them to see if they need to be taken out of service individually, writing eulogies, and making arrangements for visiting firefighters to be fed and housed. And then I have my own life going on at home. I can't totally ignore my own family.

"My husband is still in New Mexico, because I told him, 'Don't rush home.' I don't want him to come home at all right now, because I'm so busy. I'm hardly ever home, and when I am, it's really a pleasure to not have to talk about it anymore. I'd like to be able to not answer the phone if I don't want to, because at work I don't have any choice in the matter. I can't deal with one more person who needs me right now.

"The firehouse is unbelievably busy. It's nonstop. Before this, I couldn't imagine that I'd be in the firehouse for 24 hours and literally have no downtime. Usually, at some point during a 24-hour tour, you're doing something for yourself. You see a show on TV or you read a book. This is not happening. It's still not over. I don't have a minute to myself.

"There are still phone calls and people coming by. When they opened the streets back up to vehicle traffic, it was like nonstop tourists coming by, wanting to talk to somebody. They wanted to hear about the guys. We have a memorial set up out front with their fire department ID photos, and people come by and want to know, 'Was this guy married? How many children did he have?' It takes up a lot of time to do that, to talk to people, but we did want to tell their stories.

"I've never been one to wear fire department T-shirts off duty. I'm just not a T-shirt girl to begin with. Plus, when you're a woman in the fire department and people see you with a shirt on, they think you're somebody's wife. 'What company does your husband work in, dear?' And I have to tell them, 'No, it's not my husband; it's me.' Then it stirs up that whole 'Do women really belong in the fire department?' thing. I just hate it.

"So for 19 years, I didn't wear anything related to the fire department. But after September 11, I couldn't stop wearing that stuff. Every day I had something on with a fire department logo. In my 19 years in the fire department, I have never been more proud to be a firefighter. I couldn't even say, 'I've never been more proud to be a firefighter' without crying.

"When I was out in New Mexico, I was concerned because I couldn't get in touch with the other women on the FDNY. There are four women assigned to the 1st Division in Manhattan, and another four women assigned to the 11th Division, which is right over the Brooklyn Bridge. All of them could have been at the scene before the buildings came down. That's at least eight women who were close possibilities. Then I found out that Maureen McArdle-Schulman's company came down from Harlem, so there were nine women who easily could have been lost.

"There are only 25 of us, and 7 women were housed in companies that lost a total of 48 firefighters. We came extremely close to wiping out a third of the women on this job. It's amazing that we lost no women firefighters. But we've definitely been going through it, and we've all quietly been here for each other.

"We're here, and we've been here for 19 years in New York City, and even longer in other parts of the country. We're doing everything that every guy is doing, and sometimes we're doing more. A guy will come in and share things with me that he's not going to share in the kitchen with the other guys for fear of what they'll think. Guys will come into my office and cry to me who aren't going into the office and crying to my lieutenants. As women, I think we've got a big burden. The burden has always been on the women, because we're in a fishbowl.

"And that certainly hasn't changed now. I'm sure people are waiting to see how I'm reacting and how I'm going to handle this, as they are with other women throughout the whole city. I think it's important for people to know that we're here, that we do this job, and that we face the same fears and the same dangers as everybody else. Sometimes we even handle the situations better.

"When I took Engine 4 a year ago, I thought I'd only spend two years in Manhattan. I liked being in Brooklyn, rolling around and going to fires. It wasn't a terribly busy company, but we saw enough work, and it was fun. I worked next to the neighborhood where I grew up, so it was even that much more fun. I figured I'd be here a few years, then I'd worm my way back to Brooklyn.

"Now I know that's not going to happen. When I came back here after September 11, I had a very specific goal. I knew I had to be strong. How would it look if the captain of the company said, 'Oh, I can't work in this firehouse anymore?' Never in my 19 years in the fire service has my job been clearer to me than it is now.

"I have to stay here to pick up the pieces. I have to stay here to show a whole bunch of young kids that you don't quit when something bad happens. I have to stay here to remind them, 'You will survive this, you will get through this, and we will be a company again.'"

SARAH HALLETT, Ph.D.

AGE 38
POLICE OFFICER/POLICE PSYCHOLOGIST
ORO VALLEY POLICE DEPARTMENT
ORO VALLEY, ARIZONA

Sarah Hallett, Ph.D., is a police officer committed to helping her own. As both a patrol officer and a police psychologist, she uses her knowledge and experience to help fellow cops, firefighters, and EMS workers through some of their toughest emotional battles.

Sarah did her doctoral dissertation on trauma and stress in the homicide, child abuse, and undercover units. She worked as a staff psychologist at the federal prison in Tucson, then began consulting with law enforcement agencies. She later decided to enroll in the Police Academy.

"I felt that becoming a police officer myself would assist me in my work as a police psychologist," she explained. "I wanted to see the world through the eyes of those I was trying to help."

Although her home department is in Arizona, Sarah has spent countless weeks in New York City since the terrorist attacks on September 11, assisting local agencies that provide counseling and peer support services to rescue workers. She also served as support staff with Concerns of Police Survivors at National Police Week in Washington, D.C., and plans to return to New York to provide additional support during the first anniversary of the tragedy.

"Unfortunately, much of the country has turned its focus to bioterrorism and the military, thinking that things at the World Trade Center are over. But for us, it's just the beginning. The emotional fallout is going to be tremendous."

—Sarah Hallett, Ph.D., November 9, 2001

"I've always been interested in criminal justice, so I went into the Army for the college fund and completed my Bachelor's and Master's degrees in forensic psychology. While I was in the military, I became certified as an emergency medical technician and was a part-time firefighter in Kentucky, where I was stationed. That was my first exposure to emergency services work.

"I've been a psychologist since the mid-1990s and a police officer since 1999. I work in Oro Valley, Arizona, where I'm a patrol officer and a hostage negotiator. I'm also the clinical director for our Critical Incident Stress Management (CISM) team, which provides a series of intervention and support services to our own in law enforcement, emergency medical services, and the fire service. We also have outreach programs to assist family members following an incident that overwhelms their normal coping skills.

"I work primarily with law enforcement, but a lot of my work also involves fire and EMS, because we debrief major incidents together. It helps a lot having had that brief exposure as a firefighter and EMT, because I can draw on that experience to help others. My whole life has been in public service, which I see as a way of giving back to my community.

"Police, fire, and EMS personnel are the toughest, most resilient people we know. They wouldn't survive their careers much more than a year if they weren't. But that doesn't mean some incidents don't hit home in a real, personal way. If someone in emergency services is seriously injured or killed, or in this case, if there's a disaster on the magnitude of the World Trade Center, we provide support services and assist them in getting back to duty, being operational, and staying healthy.

"After September 11, I was on standby to go to New York City for a number of law enforcement agencies. Then one of the police chaplains from the Arizona Highway Patrol had contact with an NYPD peer support organization, Police

Organization Providing Peer Assistance, and the director of that program, Bill Genet, asked me to come out and assist. I was in New York for seven of the first ten weeks after the attack.

"When I first got there, I helped organize response teams and provided crisis intervention services. Even though these services were designed for police officers, we were seeing people in EMS, fire, and even the construction folks. We didn't turn anyone away. We realized that we could have problems of epidemic proportions if we didn't get in early, provide a safety net, and intervene where we could.

"Even after the initial response on September 11, rescue workers in New York City were working 12- and 14-hour days. For a while they weren't getting any days off, and when they did, some of them came back on their own time to help with the recovery efforts. So they were dealing with a lot of fatigue and the physical impact of this kind of work. Of course, a lot of them were going to memorial and funeral services on top of all that, then returning to Ground Zero to dig some more.

"When you're dealing with civilians who've been victims of a traumatic event, their reactions come right away. They start dealing with their feelings immediately and with even more intensity over the next several days. But there's a different timeline for emergency personnel. They tend to stay up and operational, and there's a lot of adrenaline in the rescue effort. We call this the 'heroic phase.' There's a lot of public support for emergency services personnel and the heroic efforts they made on September 11, and there's still a lot of activity going on at Ground Zero, the landfill, and the morgue. There isn't a lot of family time and no time to discuss their reactions to all this.

"There are also increased security concerns, which means that people are deployed all over the country to increase national security. When the anthrax scares and hoaxes and actual incidents occurred after September 11, it further taxed police, fire, EMS, and hazardous material operations. They had to respond to protect people, secure buildings, investigate threats, and deal with public panic, hysteria, and fear. Some of the larger cities, like New York City, had to respond to as many as 100 bioterrorism calls a day on a 12-hour shift.

"On top of that, when the president or vice president came to visit New York, they had all those extra security measures to provide. At the same time, they were still responding to fires, medical emergencies, and law enforcement calls in their own precincts and jurisdictions. So it really taxed all the personnel in every branch of emergency services. I don't think the general public realized how much was going on at once.

"There is a tremendous amount of pride and dedication among these people. These are tough people who are very persistent in getting the job done. They, too, had family members, friends, brothers, and sisters in emergency services who perished there, and they had to face the grim, but honorable, task of recovering those people. Even though it's taking a toll on them, there's a sense of, 'We're going to prevail; we're going to do the right thing to honor the people who made the ultimate sacrifice.'

"Within a few months of the World Trade Center incident, there was a plane crash in the Rockaway section of Queens. This further compounded matters for emergency services personnel and triggered reactions to the September 11 attacks. Many people wondered if we were under attack again. When I went out with an NYPD officer to assess the situation and prepare a CISM response, many of the same people I'd seen digging at Ground Zero were out there recovering bodies in Queens. That was one of the saddest things I've ever seen.

"Usually when people come off a major incident, they come to a central location and debrief from that tour. We generally take time to go over the normal reactions that they might be having and give them some simple tools for stress management. Then they grab something to eat and go get some rest.

"But everything was different at Ground Zero. They were working long hours and not coming back to a central location. Many didn't even go home. They went to Red Cross respite centers and different places that gave up bed space, or they slept a few hours in their vehicles and then went back out to dig with the bucket brigade.

"So we took our work to the field in a more informal, proactive way. For weeks we went out in small teams to check on the welfare of people in the field, both down at Ground Zero and at the temporary morgue at Bellevue Hospital. The teams consisted of police officers, paramedics, and firefighters, all trained in peer support and CISM. We would go down to Ground Zero and find people, whether they were standing on a perimeter post or taking a break, and informally check on how they were doing. We'd provide educational material about the effects of stress on emergency workers and give out hot-line numbers and counseling locations. We weren't trying to dredge up a lot of feelings. Our only goal at that time was to help keep people operational.

"We also set up a similar operation at the landfill on Staten Island, where hundreds of detectives and others were sifting through the rubble every day, looking for remains and identification. That's a pretty gruesome task. We put up a trailer on top of the hill, where, if somebody needed to talk, they could come in for support and assistance. We also provided basic information and visited informally with people when they went into the mess hall for their meals. Each time they came on duty at the landfill, we gave them an overview of what our program and services could provide.

"That's still going on. People will be working at the landfill and the morgue for up to a year. Even though they rotate in and out of those locations, while they're there, they're immersed in death and destruction. It's a very unnatural thing for anyone to do. We want to educate them about that and reassure them that they're not going crazy if they're having images of what they've experienced. What's crazy is what happened on September 11. We try to help them keep that perspective.

"What makes this incident even more difficult is the fact that it involved human-induced evil. It wasn't a natural event, like a hurricane or an earthquake. Even though those kinds of events can be tragic and can take many lives, it's much more difficult for people, psychologically, to swallow a situation like this.

"Initially, people were extremely exhausted. In some cases, they had that '1,000-yard stare.' They were very tired, yet very determined to continue. A lot of people discussed their feelings of loss in terms of the people they knew who had perished, and obviously, there was anger about the fact that this had happened at all. Many of them were dealing with the stress of trying to manage work and family life. A lot of people didn't see their families for days, and I heard that some of them didn't go home for weeks. When they did go home, of course their loved ones were terrified that something would happen to them, worried about their welfare, and sometimes didn't know how to support the person who was out there working every day.

"For people with children, especially young children, the saddest thing for me was to hear them talking about these little kids pulling on Mom or Dad's pant leg, not wanting them to go back to work. They hadn't seen much of that parent, and they were terrified. Whichever parent was left at home was juggling a lot of tasks, maybe working at an outside job, taking care of the children, managing the bills, handling everything that must be done to keep a home functioning while the emergency services person was out there working day after day. It can cause a lot of stress and strain on everyone. It bleeds over into the family, whether we want it to or not.

"Some of the preliminary estimates from the aftermath of the Oklahoma City bombing in the mid-1990s suggest that there has been a 300 percent increase in the divorce rate among rescue worker families in that city. That operation was pretty much completed within a few months, and we're really just beginning here.

"Providing family services is very important, but the magnitude is huge. There are 33,000 police officers in the NYPD, another 1,350 in the Port Authority Police Department, 11,500 firefighters, and thousands working in EMS. How many families and loved ones do they represent? The families are the unspoken heroes behind the scenes, supporting their spouses, parents, siblings, and partners who are working out there every day. So it's a real concern.

"At this point in time, our rescue workers are starting to process things more, both mentally and emotionally. We're moving into more formalized services and structured discussion groups, enabling people to vent their reactions and giving them more specific information about how to cope. We're giving them stress management tools and ideas for assisting with the family—those sorts of things. All our work is grounded in the guidelines of the International Critical Incident Stress Foundation.

"It's very important that we get in early and provide a spectrum of support services and interventions in order to help reduce the number of people who might potentially develop full-blown post-traumatic stress disorder (PTSD). This is a disorder people experience when they've been exposed to an extreme traumatic event or incident that's very overwhelming. It's similar to what many Vietnam vets and other veterans have experienced.

"One of the things that can happen is referred to as a 'flashback.' That's a replay of the event in the mind. These images get etched into people's minds, just like a videotape. It might be the image of a plane coming into a building, the buildings collapsing, or the sight of so many casualties. If this goes on for an extended period of time, it can be very debilitating. The rescue worker can start avoiding certain things, like not wanting to come into the city, not wanting to be near tall buildings, and, like many others in the general public, they may have strong startle reactions and become hypervigilant.

"There can also be extreme tension, adrenaline surges, and increased heart rate at the sound of a loud noise. Some people have a numbing of emotions, difficulty sleeping, or nightmares. Just like the videotape in the mind during waking hours, nightmares are another way the mind is trying to process some very unusual and abnormal images during sleep. There is also a possibility that some people will experience depression, anxiety, and substance abuse.

"The tragedy here is that, for public safety and emergency services personnel, the casualty rate from September 11 could become much higher through suicide and other potentially debilitating circumstances. Statistically, police officers are much more likely to kill themselves than be killed in the line of duty. What we want to prevent are further casualties from September 11.

"Our nation has been encouraged to move on, but it's the emergency services workers who will be left to live with this in a whole different way than the general public. Unfortunately, much of the country has turned its focus to bioterrorism and the military, thinking that things at the World Trade Center are over. But for us, it's just the beginning. The emotional fallout is going to be tremendous.

"Most people remember the bombing in Oklahoma City and the image of the firefighter carrying a baby from the scene. That picture is etched in our minds from that day. But what most people don't know is that the firefighter in that picture tragically took his own life. And the rescuer who saved baby Jessica from the well in Midland, Texas. We all remember that. But did you know that he killed himself, too? These people can do the most heroic things, make the most heroic attempts to rescue people, but they're left to live with some really horrendous things. It's not a question of toughness. These things really hit home with people, and we've got to provide long-term support to help them.

"We learned our lesson the hard way with Vietnam. We didn't provide those kinds of services, and the nation turned its back on those vets. It's difficult to collect statistics, but there were something like 57,000 combat casualties in Vietnam, and the subsequent suicide rate has been even higher than that. More people have killed themselves after coming home from the war than died in the war itself. Emergency services workers will tell you that, even today, we still get calls to respond to suicidal subjects who are Vietnam vets. These people continue to relive some pretty horrific things.

"In the emergency services field, sometimes we're so busy helping our communities that we don't stop to help each other. I think that the culture in law enforcement, fire, and EMS can sometimes get in the way of encouraging people to talk openly about their reactions to events that have turned their world upside down. Many people have a big resistance to getting help, because there's a lot of stigma attached. *What are people going to think?*

"I, too, have been working 12 to 14 hours a day at Ground Zero or the landfill or the morgue, and so have many other CISM team members. We recognize that we've been exposed to things that the average citizen normally wouldn't see, and even as we're helping the helpers, we can be affected, too. We tend to be really good at checking on the welfare of strangers all day, but we don't always hit the pause button and take the time to check in with ourselves. That's why we have a program in place to debrief the debriefers at the end of each day. When teams rotate back home to whatever state they're from, they're also encouraged to have a separate debriefing to process their reactions with an objective person.

"One of the recent changes we've made in our occupation is that we've trained police officers, firefighters, and paramedics to be peer support members of the mental health team. Even though we use mental health personnel to support and assist, it's a lot easier for a firefighter to talk to a fellow firefighter than to a shrink. These are people who really care and who want to help their peers.

"There's also an increased emphasis on confidentiality. That's another reason why we're rotating volunteer teams into New York City from out of state to help support the peer program. Our presence has gone a long way to support and maintain the morale of local teams who are out there every day. We're not going to forget, and we're going to continue to provide support, because we really care about them.

"All these people who come in from other states are volunteers. In my case, my chief understands the importance and value of providing these services, and he supports my being there. Others have covered my practice as a police psychologist, which is great. Most of the people responding to New York City are doing so with the support of their home departments.

"As human beings, no one is immune to the effects of what happened on that day. I know this will be life-changing for me, yet it's probably some of the most important work I've ever done. I believe that time will tell what the full impact will be on any of us. One of the things I've noticed already is how a lot of things that used to be important to me seem insignificant or inconsequential now. I've seen so much death and destruction that a lot of other things just don't seem to matter.

"I also see a lot of unity and acts of pure human kindness that give me faith in the future. I believe there are always lessons to be learned in everything we do. One of the lessons I've learned from this is that, despite the most challenging circumstances, there are good people out there who are always ready to help us."

IN REMEMBRANCE

October 21, 1976–September 11, 2001

YAMEL MERINO

AGE 24
EMERGENCY MEDICAL TECHNICIAN
METROCARE AMBULANCE SERVICE
MT. VERNON, NEW YORK

When Yamel Merino was caught in the collapse of the South Tower of the World Trade Center, she became, at 24, the youngest female rescue worker to lose her life on September 11.

An emergency medical technician with MetroCare Ambulance Service in Mt. Vernon, New York, Yamel was loved by patients and colleagues alike. Yamel and Sande Santiago—inseparable friends and co-workers—took special care of their elderly patients. While waiting at a nursing home for the discharge paperwork necessary for transfer, they often entertained the frail or ailing person in their care by grabbing a pencil or hairbrush to use as a microphone, and singing and dancing along to a song on the radio.

We spoke with Sande about Yamel in April 2002, and with Yamel's parents, Leslie and Anna Jager; her sister, Gabriella Sierra; and the vice president of MetroCare, Jim O'Connor. The constant refrain we heard was, "It sounds like a cliché, but Yamel really did light up a room. When she walked in, you knew she was there. She always had a smile and a kind word for others. She was one of a kind."

The love and affection Yamel inspired in all who knew her was summed up very simply by her best friend Sande, who said at the end of our interview, "She was an angel."

Yamel's eight-year-old son, Kevin, and the rest of her close-knit family and friends, including seven brothers and sisters, are struggling to come to grips with the loss of this beautiful young woman who adored her son and had a future filled with promise.

"I want to see Kevin grow up," Yamel said shortly before the attacks on the World Trade Center. "I want to live forever."

"I'm not complete anymore. Even though I have my other children and I love them, I feel lonely for Yamel."

—Anna Jager, April 28, 2002

On September 11, Yamel Merino responded without hesitation to the call of duty at the World Trade Center, and died while trying to save others. She was assigned that day to a 911 ambulance and was working in a triage area at the Marriott Hotel, across the street from the Twin Towers, when the first Tower collapsed.

When co-worker and friend Sande Santiago first heard about the attack, she called Yamel on her cell phone. Yamel had not heard the news yet, but the order came over the radio in her ambulance as she spoke with Sande: "15 George, respond to the incident at the World Trade Center." Yamel told Sande she'd give her a call when she got downtown. Sande waited and waited and waited for Yamel's call.

"I knew something had happened," she says, "because Yamel was really responsible, always keeping to the rules and regulations of the 911 system. She was supposed to give updated signals every 20 minutes. I knew something was wrong when I never once heard her come over the frequency. And not contacting her mother—that would have been number one. She would have called her mother if she could."

On the day of the attacks, Yamel's mother, Anna Jager, sat on her front steps, shocked and terrified, waiting for her daughter. Sande knew that Mrs. Jager would sit on those steps until someone called her with news about Yamel or until Yamel returned home. MetroCare Supervisor Elsie Torres, who was like a second mother to Sande and Yamel at work, accompanied Sande and a counselor to Yamel's apartment. Sande broke the news to Mrs. Jager: "Yamel is missing."

"On Tuesday night, I gave the family my home phone, work phone, and cell phone numbers," says MetroCare Vice President Jim O'Connor. "Yamel's father called at 1:30 A.M. and simply said, 'Mr. O'Connor, have you found my daughter yet?' One of the hardest things I've ever had to do was tell him, 'No.'"

The following day, Yamel's family and friends gathered at the home of her parents, crying, praying, and desperately hoping Yamel would be found alive. Jim, the father of a daughter and two sons very close in age to Yamel, sat in his car several blocks away, waiting for the call he dreaded. He did not want Yamel's family to receive any information from a stranger, from someone who did not know and love their daughter.

Jim describes this as the hardest day of his life.

"The call finally came, and I had to tell Yamel's family that her death had been confirmed," he says. "When I walked into the house, I was praying that I wouldn't have to say the words out loud, but they all just looked at me and were totally quiet. They were all staring at me, not saying a word. I had to tell them, 'We found Yamel, but unfortunately, she wasn't able to get out.'

"Then people just started screaming. Everyone was crying, and her sister was saying, 'I don't believe you! I don't believe you! You have to take us there! I want to see her!' Her mother was saying that she wanted to go to her. I took Mr. Jager outside and said, 'I'm telling you, as the father of a daughter—you don't want to go down there.' Then Mrs. Jager came out and I said, 'I'm begging you; let me take it from here. I promise you I will bring your daughter home.'"

Yamel's parents trusted Jim and allowed him to make the necessary arrangements. "Through it all," he says, "my heart was breaking for them."

"That day was so horrible," says Yamel's sister, Gabriella Sierra. "I remember it like I was in a dream. When Jim walked in the house, it felt like one of those scenes in a movie, where everything slows down and you only see one thing. I went right up to him, and he took my hands and looked into my eyes. All I could see was Jim's face, and his voice sounded like it was echoing.

"When he said Yamel died in a violent explosion," Gabriella says, "everyone was crying, and my mother was screaming, 'I don't want to hear it! I don't believe you!' I was saying, 'You're a liar! How do you know it's Yamel, if it was a violent explosion? I want to see her.'

"I had to go to the funeral home myself, because I could not bring myself to believe that she was gone," she adds. "They pulled the sheet back, and I could see right away that it was her. I gave her so many kisses. I kissed her for her son, Kevin, for her friend, Sande, and for everyone in my family. That was the hardest thing, kissing my sister good-bye."

When Mrs. Jager talks about Yamel, her words are punctuated by deep sobs. She describes the devastation of living with the loss of her beloved daughter. It gets worse with each passing day, she says, not better, as people keep telling her it will.

"I'm not complete anymore," she says. "Even though I have my other children and I love them, I feel lonely for Yamel. We were so close. We were in touch two or three or four times a day. She would call just to say, 'Hello Mother, how are you doing?' She would drop Kevin off at 6 in the morning and then pick him up after work. Even on her days off, she'd come over and stay at my house. I'd say, 'Why don't you go to your own apartment?' But she'd say, 'Oh Mother, I want to be here with you.'"

Yamel's mother remembers the last time she spoke to her daughter. "On that morning, she called and said, 'Mommy, I forgot to give Kevin lunch money.' We talked for a minute and then—this is our way of saying good-bye—she asked in Spanish for a blessing, '*La Bendición, Mommy.*' I said, '*Que dios te bendiga,*' which means, 'May God bless you.'

"Lots of mothers will tell you that their daughters are special," she adds, "and people will say, 'Oh, that's just the mother talking.' But in this case, it's really true. There was something about Yamel—anyone will tell you this—that was so wonderful, so wonderful. I can't even describe it."

The loss of Yamel was all the more painful to those who knew her because she had worked so hard to create a good life for herself and her son. Everything was just beginning to come together for them. As a young mother responsible for eight-year-old Kevin, it had not been easy for her to become an EMT, save money, and maintain a full-time job. She was proud to have moved into her own apartment with a nice room for her son and to have recently bought a new car.

Yamel began working at MetroCare in June 1997, when she was 20 years old. "She loved her job," her mother says, "and she always came home with a story to tell." Yamel told her family stories of helping to deliver babies, working with elderly patients, and facing new challenges every day. In 1999, the United New York Ambulance Association named her EMT of the Year.

"Yamel was an employee you wished you could clone," says Jim. "She made every patient she took care of feel special. Whether it was an emergency call for a heart attack or transporting an elderly person to a nursing home, she treated all her patients as if they were related to her. They would call MetroCare and say, 'You need to send that girl back here.' They would ask for her by name. If you were sick and wanted someone to give you special attention, you'd want Yamel there with you."

In addition to being a hard worker, Yamel was also fun-loving and funny. "She loved to do imitations," Sande says. "She'd come in first thing in the morning, before people were even awake, and do that pose from the woman on *Saturday Night Live.* She'd jump down on one knee, throw her arms up, and say, 'Superstar!' Or she'd imitate Jim Carrey. She loved that movie *Dumb and Dumber,* and she knew so many parts by heart. She always had me cracking up with that."

Yamel and Sande made quite a team. They enjoyed entertaining their elderly patients, particularly those suffering from Alzheimer's disease or dementia. "We knew that people thought they couldn't understand anything," says Sande, "but we always felt that they could sense we were there and that we cared about them."

The two friends loved to laugh together. As they drove around, they played "Punch Buggy," a game Yamel and Gabriella had played since high school. The first to spot a Volkswagen Bug would punch the other in the arm while calling out,

"Punch Buggy!" "But we had to stop doing that," says Sande, "when the new VW Bugs came out. They were all over the place, and we were both black and blue."

Sande's eight-year-old son, Mathew, and Yamel's son, Kevin, became good friends, which cemented the already close relationship between the two young mothers. Sande remembers Yamel's love and devotion to Kevin. "He was everything to her," she says. "She was such a good mother, so loving. Everything she did was for Kevin. She wanted him to be happy and comfortable. She made sure he earned what he got, and she always saw to it that he had what he needed."

Sande and her husband offered to take Yamel and Kevin with them to Disney World. They knew Kevin wanted to go, and they thought it would be a great time for the boys. Knowing Yamel couldn't afford it, they offered to cover her expenses. Yamel declined, saying, "I want to do this for Kevin. I want to take him to Disney World on my own." And she did, in July 2001, just two months before the attack on the World Trade Center. The photographs of Kevin and Yamel during that trip—standing under a huge globe bearing the word *Universe,* riding on a boat, posing near a statue—show a mother and son having the time of their lives.

Kevin, who now lives with Yamel's parents, does not understand why his mother was taken from him. When his grandfather sat Kevin down and broke the news, "He just walked away," says Gabriella. "He said, 'I don't want to talk about it.' Since then, he has said, 'Why did my mother have to be there?' Sometimes he talks about how much he hates those people who did that to the World Trade Center."

Yamel's mother simply says, "He wants Mommy to be home, and he's angry that she's gone."

Gabriella says that Yamel's friends always wanted to spend more time with her. "Everyone wanted her friendship," she says, "because she was such a good person. Her goodness always shined through. She was the life of everything, always warm and affectionate and funny. She had the warmest smile, the best hugs, and even when she was alive, she was always really missed when she wasn't around.

"There were so many people at her funeral," she adds. "It looked like a demonstration. A 15-minute ride to the cemetery took more than an hour because there were so many cars."

Gabriella offers a word of advice: "Hold on to your memories. Take advantage of every moment you have with the people you love."

Gabriella is pregnant with her second child. Anticipating the arrival of this baby is especially poignant for Yamel's family, as the baby is due to be born close to September 11.

If Gabriella has a girl, she plans to name her "Yamel."

Because Yamel was employed by a private company and wasn't considered a municipal employee, she was not eligible for the same funds as FDNY and NYPD employees. A trust fund has been established for Yamel's son, Kevin. Donations may be made to The Yamel Merino Memorial Trust Fund, PO Box 1708, White Plains, NY 10602.

May 13, 1955–September 11, 2001

CAPTAIN KATHY MAZZA

AGE 46
COMMANDING OFFICER OF THE
PORT AUTHORITY POLICE ACADEMY
PORT AUTHORITY OF NEW YORK
AND NEW JERSEY POLICE DEPARTMENT

We first attempted to reach Captain Kathy Mazza's husband, Chris Delosh, only six weeks after the attack on the World Trade Center. When we called the home he had shared with Kathy in Farmingdale, New York, it was devastating to hear Kathy's voice on the answering machine, cheerfully inviting callers to leave a message. We realized it was too soon to approach Chris, an NYPD police officer in New York's 25th Precinct in Harlem. We waited until April 2002 to speak with him about Kathy.

We also spoke to Kathy's friends and colleagues, Peter Killeen, M.S., a police psychotherapist, and Jessica Gotthold, a special agent with the Bureau of Alcohol, Tobacco and Firearms.

The pride with which Chris, Peter, and Jessica described Kathy painted a vivid portrait of a woman with an incredible gift for helping others and making things happen.

Kathy's husband told us that she went into police work because her previous career as an operating-room nurse with St. Francis Hospital's elite cardio-thoracic team had become routine, and she needed new challenges. She rose quickly through the ranks of the Port Authority Police Department, and everyone who knew Kathy believed she would one day achieve the rank of chief.

"Kathy always told me that she had to work so hard because she was a woman," said Chris. "She wanted to leave a mark on this world, and she definitely did. She saved lives—what more can you ask for? That's what Kathy was all about. Helping."

"Kathy affected so many people. She definitely did leave her mark. She is a part of history, and she will be remembered."

—Chris Delosh, April 25, 2002

Captain Kathy Mazza, the first woman in the Port Authority Police Department to die in the line of duty, was, by all accounts, a person who made a deep impression on others and who made things happen wherever she went. Her determination, kindness, strength, and sense of humor emerge from the stories told by those who loved her.

"She was a dynamic woman and a real role model for me," says friend and colleague Jessica Gotthold, a special agent with the Bureau of Alcohol, Tobacco and Firearms. "She was an amazing achiever on every front."

Before the attack on the World Trade Center, Kathy and her husband, NYPD Police Officer Chris Delosh, had been looking forward to celebrating their sixteenth wedding anniversary. Chris speaks of Kathy not only with sorrow, but also with a deep sense of pride in who she was and how she lived her life.

"Kathy always had the mind-set, 'If you can do it, I can do it,'" says Chris. "Nothing would hold her back. A few years ago, a tree fell on our property, and it needed to be cut up with a chainsaw. It was taking me a few days to finish the job, and Kathy said, 'Hey, show me how to work this chainsaw. I want to finish this up.' So I showed her, and she went at it, chopping up the rest of the tree. That's how she did everything.

"Kathy was also great with kids," adds Chris. "We weren't blessed with children of our own, but she had such good relationships with her nieces and nephews, even when they became teenagers and young adults. They always sought her out, and she was very close to them. She was also thoughtful toward new parents, giving them a break by going to their house and saying, 'I'll take care of the baby; you take off for a while.' She was always so great that way."

All who knew Kathy describe her as a person with a great passion for helping others, so it was no surprise that people constantly turned to her for advice. She was quick to help others on the job, in her family, and in her neighborhood.

Rushing a neighbor in the beginning stages of a heart attack to the hospital, tending to a young boy next door who'd cut himself, going beyond the call of duty to help a homeless person, and getting her own mother correctly diagnosed when she was having chest pains—this was life as usual for Kathy Mazza.

On the morning of September 11, Kathy was at the Port Authority's Police Academy in Jersey City, conducting interviews with prospective instructors. Kathy and her colleagues had just received the okay to move the Academy from Seagirt, New Jersey, to the Port Authority's Tech Center in Jersey City, a stone's throw from the Holland Tunnel and only a few minutes away from the World Trade Center. When the call came in that a plane had crashed into one of the Towers, Kathy headed over without a moment's hesitation.

"I called her office and she was already gone," recalls Chris. "I assumed that she was going to set up a temporary Command Center, so I felt a little bit secure that that's what she was doing. I beeped her, thinking she'd call me when she could, when she had the time. I found out later that because it was so chaotic, there was no time to set up a Command Center. She went directly into the building and started to evacuate people."

Chris, an 18-year veteran of the New York City Police Department who is not easily shaken, began to worry.

"The Port Authority radios weren't working at that point," he says. "Around that time, the first building collapsed, and there was a call for everyone to evacuate. I hadn't heard from Kathy in a couple of hours. I really started to worry. I wanted to at least know if someone knew where she was."

Several people reported seeing Kathy in action in Tower One that day. At one point, the revolving doors in the lobby of the building were jammed with panicked people trying to push their way out. Kathy calmly shot out several huge plate-glass windows, enabling hundreds of people to escape.

Based on eyewitness accounts, it is believed that Kathy made it to the twenty-ninth floor. She helped evacuate and save the lives of thousands of terrified people attempting to flee Tower One.

Chris held out hope that Kathy might be found alive. He, like many others, recalled the Oklahoma City tragedy, when people were still being found alive weeks after the bombing. Chris, along with the rest of the country, would soon learn that no one was likely to be found alive among the ruins of the World Trade Center.

Chris says that Kathy wanted to leave a mark on this world. "I've heard from her former teachers, patients, doctors, and co-workers. Kathy affected so many people. She definitely did leave her mark. She is a part of history, and she will be remembered."

Kathy grew up in Massapequa, New York, where she was raised with three brothers. Chris suspects that this is where she developed the competitive edge that would later characterize her work in nursing and law enforcement, as well as her rapid rise through the ranks of the Port Authority Police Department.

When Kathy graduated from nursing school, she respectfully declined to be "capped" during a ceremony in which a small white nurse's cap was placed on the head of female graduates. "If men won't wear this, I won't either," she said. She eventually became an operating-room nurse with the elite cardio-thoracic team at St. Francis Hospital on Long Island.

"Kathy would never back down when it came to getting patients the care they needed," says Chris. "She would go up against doctors or anybody else. She had strong opinions about what kind of care a person might need, and she was almost always right on the money. I once asked her why she didn't become a doctor, because she was so good at it. She said, 'No, that's okay. It's time to do something else.'"

Seeking new challenges, Kathy entered the Port Authority Police Academy in 1987, graduated that same year, and was assigned to Kennedy Airport. Her background as a nurse and EMT influenced her police work, and she was responsible for saving a number of lives by introducing defibrillator training at New York airports. Kathy saw to it that 600 officers were trained on defibrillators, and at least 17 heart-attack victims have been saved as a direct result of her vision and determination.

In April 1999, Kathy made captain and later became the first female Commanding Officer of the Port Authority Police Academy. That same year, she was named New York City's Basic Life Support Provider of the Year, one of the many times she was recognized for going above and beyond the call of duty.

Jessica Gotthold met Kathy in 1995 when they were both teaching at the Academy. Jessica laughs as she recounts her first meeting with Kathy:

"I had just started teaching at the Academy, so I was all dressed up. I was wearing a nice skirt and jacket, and I heard this voice behind me say, 'Hey, nice legs—do you power lift?' It was so funny. She loved that little bit of shock value."

Jessica speaks admiringly of her friend. "She moved up fast, and deservedly so. She had just made sergeant when I met her. If you didn't know who or how she was and you heard that she'd moved up from patrol officer to captain within six years, you'd say, 'Wow, she must have had some kind of hook.' But in her case, it was merited. It was because she was unstoppable.

"Kathy was such a shining star," adds Jessica. "They couldn't deny her, and she wasn't going to back down. She wasn't going to go away quietly and be satisfied with being passed over. She knew that as a female in law enforcement, you have to give at least 150 percent to gain the respect and acknowledgment of your colleagues. Kathy was always willing to do that. She worked incredibly hard, and she earned every promotion."

As Commanding Officer of the Police Academy, Kathy was one of the highest-ranking women in the Port Authority Police Department. Her achievements were the result of a lifetime of giving more than was required or expected of her. It seems that all the events of her life brought her to that moment on September 11 when she ran into Tower One at the World Trade Center to help evacuate thousands of terrified people.

Kathy's body was finally recovered on February 9, 2002. She was found in the lobby of Tower One with five co-workers from the Port Authority surrounding a woman in a wheelchair. It is believed that they were helping the woman escape the building.

"Part of Kathy's makeup was to be a caretaker," says Peter Killeen, a police psychotherapist and colleague of Kathy's. "That quality brought her to where she was that day. I don't think she would have done anything differently, maybe not even if she knew the building was going to come down. She was just one of those people. She was committed as an officer and as a human being. It didn't matter who it was, she would have been there to get them to safety. She was truly a hero who died in the line of duty."

Kathy was buried near her maternal grandmother, with whom she had been very close. "Once they recovered her body," says Chris, "I thought, *What do I do now?* I don't like the word 'closure.' I don't believe that there is any closure. I didn't go to Ground Zero. I don't want that to be my final memory of Kathy. That's not the way her life was supposed to end."

"There is a sanctity in death," says Peter, "and it's both scary and beautiful. There is no question that Kathy and the others have a place in heaven. They've certainly earned their entry."

Kathy traveled a path of her own making, helping others and saving lives at every step along the way. She always used to say she'd retire as a police chief. Those who loved her know that the achievements of her life and her courageous final hours elevated her to an even higher rank.

She will be remembered as Kathy Mazza, hero.

February 14, 1963–September 11, 2001

MOIRA SMITH

AGE 38
POLICE OFFICER
NEW YORK CITY POLICE DEPARTMENT
13TH PRECINCT

When it was learned that Police Officer Moira Smith had run directly into the burning South Tower to save lives, it did not come as a surprise to anyone who knew her. We heard story after story of her courage and fearlessness. Her husband, Police Officer Jim Smith, told us that Moira never hesitated to do her duty.

NYPD Sergeant Mary Young told us of being on patrol with Moira and taking a break to grab some pizza for lunch. "I'm waiting in the car, and all of a sudden I see Moira tearing down the street on a foot pursuit, going after this huge guy," she said. "So I started running, screaming, 'Moira! What's going on?' She yells to me as she's running, 'He's got a gun!' Before I know it, she's taking him down and taking his gun from him. After we cuffed him and put him in the radio car, I said, 'Hey, I thought we were just getting a slice here!' But that's just the way she was."

The light of Moira's life was two-and-a-half-year-old Patricia Mary, "the baby," as Moira always called her. Moira and Jim reveled in being parents to this beautiful, chubby-cheeked little girl, and theirs was a close and loving family.

NYPD Police Officer and good friend Lissa Navarra told us that Moira would open her locker at work to reveal photos of Patricia Mary and Jim. "She would say to everyone within earshot, 'Look at my little pumpkin. Isn't she the cutest thing?' She would blow kisses to the pictures of the baby and her husband, Jimmy. She was always telling me about some nice thing Jimmy had done for her. 'Oh, Jimmy did this for me, or Jimmy did that. Isn't he just the best?'"

Moira loved her husband, cherished her daughter, and was passionate about being a New York City police officer. She was full of life, and everyone who knew her commented on how much she had to give.

"What Moira did on September 11 wasn't a one-time thing. That's who Moira was. That was what she did every day. She wasn't reckless, but she never backed down."

—Jim Smith, April 30, 2002

Jim Smith, an NYPD police officer, first met his wife, Moira, in 1988, when she joined the New York City Police Department. They quickly became friends and began traveling and spending time together.

An avid sports fan who loved the New York Yankees, Rangers, and Giants, "Moira was always up for anything," says Jim, who worked with Moira in the 13th Precinct early in their relationship. "We'd be hanging out after work, and one of us would get the idea to go to Baltimore for a Yankees game the next day. Before you knew it, there we'd be, on a train at 3:30 in the morning, headed to Baltimore. Or I'd say, 'We should go to Boston for Bloody Marys,' and we'd just take off and do it."

Marriage followed, and Moira and Jim continued to travel. Moira loved exploring new places, and some of her many trips included visits to Africa and Spain, where she ran with the bulls in 1995.

Even the birth of their baby, Patricia Mary, didn't slow them down. When friends told Moira that she and Jim would have to wait until Patricia Mary was older to resume their travels, Moira would simply laugh and say, "Have baby, will travel!" Moira and Jim bought an RV, in which they continued their adventures with Patricia Mary in tow.

Jim and Moira were best friends and constant companions who admired one another personally and professionally. "I was always impressed with the way she did her job," Jim says. "She wasn't careless, but she never hesitated. She liked being out there with people, and she enjoyed the excitement. She was a good cop, and she knew how to handle people.

"What Moira did on September 11—and this is true for all the officers—wasn't a one-time thing. That's who Moira was. That was what she did every day. She wasn't reckless, but she never backed down."

At one time in her career, Moira worked in a Street Narcotics Enforcement Unit doing drug buys in plainclothes. But she preferred to be in uniform. She had opportunities to go into the Detective Bureau, but she wanted to be out on the street with the people she was charged with protecting.

In 1991, when both Jim and Moira were with the Transit Police, Jim had just finished work and was waiting on the subway platform for the train that would take him home. "All of a sudden, I see Moira running, in uniform, at top speed," he says. "I didn't know what was going on, so I took off after her. It turned out that a train had crashed at the Union Street Station, and she was the first uniformed officer on the scene. We spent the next 12 hours pulling people out."

That act, for which both Moira and Jim were awarded the Distinguished Duty medal, revealed Moira's character, her willingness to run toward danger when necessary, and her commitment to doing her duty as a police officer.

The recipient of two Excellent Police Duty medals, Moira was viewed by co-workers and supervisors alike as a reliable, courageous, and caring police officer.

"She was so compassionate, so serious about her work," says Moira's friend, Sergeant Mary Young. "Nothing could have kept her from running into that building on September 11 to help save lives. Nothing."

According to her good friend and co-worker, Police Officer Lissa Navarra, Moira was assigned to a Community Policing Unit on the day of the World Trade Center attack. She was in a van on 6th Avenue and 17th Street, covering a demonstration. When the first plane struck, Moira immediately got on the radio, described the attack, and notified the Communications Division so that they could broadcast it. She reported that the plane looked like a 747 and that there were witnesses who were very shaken.

Moira brought the witnesses to the 13th Precinct, made sure they were safe, and then, true to her nature, jumped right back into a van headed for the World Trade Center. She could have easily stayed at the Precinct. But that wasn't Moira's way.

Several people saw Moira during the chaos of that day, and the *New York Daily News* published a photograph of her helping a bleeding man to a triage center. The man later came forward and met with Jim. He wanted Jim to know that Moira had gotten him to safety. Of the photo, Lissa says, "You can see her face—it's so intense. She's not panicked at all. She had a goal, and that was to get him where he had to be and then get back in there and do some more."

Everyone waited for word from Moira on that day, hoping against hope. Jim positioned himself at Bellevue Hospital, assuming that Moira would be brought in injured. As the days went by, however, it became horrifyingly clear to Moira's family, friends, and co-workers that no one was going to be found alive in the rubble of the Twin Towers.

After the attacks, an anonymous letter was posted on one of the many walls near Ground Zero that were filled with missing persons posters and messages to loved ones. The writer of this letter, Martin Glynn, was later identified by the NYPD, and he confirmed that he was the author:

"Since the terrorist attack on the World Trade Center," wrote Glynn, *"I have seen the face of NYPD Officer Moira Smith in my mind every day. This is to document what she was doing 15 minutes before the collapse of the South Tower, where she was assisting in the evacuation.*

"I was on the eighty-fourth floor of the South Tower when the North Tower was hit. We started going down the stairs almost immediately. The disaster seemed limited to the North Tower at that point. [When] the second plane hit the South Tower, the building lurched and the air pressure jumped. The temperature went up 10 degrees in an instant. We knew something immense had happened.

"We exited the stairs on the ground floor, directly in front of the large windows overlooking the plaza. I started to look at the scene of destruction, but there was Officer Smith looking in my eyes. She was waving her flashlight and saying, 'Don't look. Don't look. Keep moving.' She conveyed her authority and command of the situation. Her show of courage reassured the crowd. She kept the evacuation moving in an orderly manner despite the horrific scene in the plaza.

"When I got out of the building, I went to find a phone to tell my family I was all right. After making the call, I looked back and saw the Tower collapse. NYPD Officer Smith saved dozens, if not hundreds, of lives. Thank you, Moira. May God bless you."

"Moira put her life on the line that day," says Lissa. "She was really amazing. Truly a hero. If there was something to be done, she was going to do it. People might say, 'A cop died,' but they don't know the person. Moira had so much love in her and so much loyalty. She had a great sense of humor and a twinkle in her eyes. She was very, very caring. When she asked me how I was doing, she really wanted to know. When I walked away from her, I always felt better."

Nor did Moira worry about the little things. "Whenever I talked to her about things most people cry about," says Lissa, "like bills or personal problems, she'd always say the same thing: 'Life is too short.'"

Lieutenant Kim Royster of the NYPD Deputy Commissioner's Office of Public Information says that after Moira's death, plans were made to honor her memory by naming a ferry after her. When she contacted the family to let them know, they told a story that stunned her. According to family members, Moira and Jim had recently been trying to sell their RV. They were kidding around with Patricia Mary, pretending to cry because they had to sell it, and Patricia Mary said, "Don't worry, Daddy. I'll buy you another RV." Moira said, "Well, what about me?" The child replied, "Don't worry, Mommy. I'm going to buy you a boat."

Moira's body was not recovered until March 20. "She showed up on the first day of spring at 4:40 A.M.," says her friend, Mary. "I drove down to the site and carried her out as part of the honor guard with several other female officers. I carried her out with my own hands."

Mary spoke to Patricia Mary at Moira's funeral and asked, "Are you going to be a police officer one day?" Mary said the little girl responded with a statement that could have come straight from Moira herself. "Yes," replied Patricia Mary. "I'm going to run after the bad guy and get him."

Mary, who has a child close in age to Patricia Mary, has put together a time capsule for Moira's daughter so that when she is older, she will know of her mother's heroism. In it, she has placed every news article and magazine clipping about Moira, along with quilts, T-shirts donated by all the units in the police department, American flags, and letters from all over the world.

What Jim wants his daughter to know about her mother is very simple: "I want Patricia Mary to know how much Moira loved her," says Jim. "That's the most important thing—how much her mother loved her."

NYPD Police Officer Moira Smith assists an injured civilian on the morning of September
(New York Daily News/*Corey Sipkin*)

ABOUT THE AUTHORS

Susan Hagen is an award-winning writer whose career began 25 years ago as a newspaper reporter in Northern California. Her coverage of the health-care industry led to a series of positions in hospital media relations, including editor of employee publications for Kaiser Permanente, the world's largest health maintenance organization.

Over the last 15 years, Hagen has been retained as a freelance writer by more than 100 corporations and nonprofit organizations throughout the San Francisco Bay area. She also teaches classes in Writing as Spiritual Practice, helping others give voice to transformative experiences through the written and spoken word.

In 1994, Hagen became a firefighter and emergency medical technician as a way of serving her community in rural Sonoma County, California. She is currently a member of the Graton Fire Protection District, where she is a CPR instructor and assists with the Emergency Medical Services training program at Santa Rosa Junior College. Her experiences in public safety have provided a wealth of material for her creative nonfiction stories, which have appeared in a variety of literary publications.

Half the members of the Hagen's first firehouse in the town of Freestone were women. Following the September 11 terrorist attacks, she and her colleagues were acutely aware of the absence of women in media portrayals of rescue workers at the World Trade Center. Hagen combined her knowledge of the fire service with her experience as a writer to conceive of the idea for this book with fellow author Mary Carouba.

Mary Carouba is an investigative social worker for the Human Services Department in Sonoma County, California. A former child abuse investigator and expert witness in cases involving chemical dependency, she currently investigates elder abuse and serves as an advocate for victims and their families.

Carouba's career began 25 years ago with Delancey Street Foundation, Inc., an organization committed to changing the world one person at a time. A popular speaker, educator, and performance artist whose material is geared toward social and family issues, she appears internationally before a broad spectrum of audiences. While her presentations and performances are designed to educate and entertain, they also serve as an outlet for expressing her commitment to social justice and advocating for social change.

Carouba was compelled by the events of September 11 to undertake this book as a way of taking positive action in the aftermath of tragedy. As one who works with many women in the law enforcement community, she was inspired to acknowledge the women who responded to the terrorist attacks after realizing that the media presented few role models for girls in its reporting of the event.

In the course of researching and writing this book, Carouba's experience as a mental health professional created a safe harbor for women rescue workers to share their stories and express their emotions. She continues to seek new venues for raising public awareness of the contributions of women at this historic event as a way of assisting them in their recovery.

Susan Hagen and Mary Carouba may be reached for speaking engagements, seminars, or other educational opportunities, at:

Women at Ground Zero
PO Box 2641
Santa Rosa, CA 95405
www.womenatgroundzero.com